OCR

Biology

REVISION GUIDE

David Applin

Jon Pickering

A2

D0248721

OXFORD

Oxford University Press is a department of the University of Oxford. It furthers the University's objective of excellence in research, scholarship, and education by publishing worldwide in

Oxford New York Auckland Cape Town Dar es Salaam Hong Kong Karachi Kuala Lumpur Madrid Melbourne Mexico City Nairobi New Delhi Shanghai Taipei Toronto

With offices in
Argentina Austria Brazil Chile Czech Republic France Greece Guatamala Hungary Italy Japan South Korea Poland Portugal Singapore Switzerland Thailand Turkey Ukraine Vietnam

Oxford is a registered trade mark of Oxford University Press in the UK and in certain other countries

British Library Cataloguing in Publication Data
Data available

ISBN: 978-0-19913627-8

10 9 8 7 6 5 4 3 2 1

Printed by Bell and Bain Ltd., Glasgow

Acknowledgements:
Authors, editors, co-ordinators and contributors: David Applin, Ron Pickering, Ruth Holmes, Eileen Ramsden, Peter Marshall, Piers Wood, John Mullins, Sarah Ware.
Project managed by Elektra Media Ltd. Typeset by Wearset Ltd.

Paper used in the production of this book is a natural, recyclable product made from wood grown in sustainable forests. The manufacturing process conforms to the environmental regulations of the country of origin.

Contents

Introduction

I hope that this book helps you to do well in your examinations. Although these are an important goal they are not the only reason for studying Biology. I hope also that you will develop a real and lasting interest in the living world which studying Biology will help you to understand. If you enjoy using this book then all the effort of writing it will have been worthwhile.

David Applin

Getting the most out of this Revision Guide

This *Revision Guide* covers the subject content of the specification for the OCR Advanced GCE in Biology H421. It is not intended to replace textbooks and other learning resources but rather to provide succinct coverage of all aspects of the OCR specification, A2 Biology.

It is written to be accessible to all students. You will find it useful whether you wish to continue studying Biology after A2 or want to achieve as high a grade as possible at A2, as part of a suite of qualifications suitable for continuing studies in other areas.

This book is not a traditional textbook. Where appropriate, content is presented as annotated diagrams, flow charts and graphics integrated with easy-to-read text which nevertheless maintains standards of accuracy and scientific literacy which will enable you to obtain the highest grade possible in your exams. It includes the latest developments in the biological sciences.

Revising successfully

To be successful in A2 level Biology you must be able to:
- recall information
- apply your knowledge to new and unfamiliar situations
- carry out precise and accurate experimental work
- interpret and analyse your own experimental data and that of others

Careful revision will enable you to perform at your best in your examinations. Work with determination and tackle the course in small chunks. Make sure that you are active in your revision – just reading information is not enough.

Useful strategies for revision

As well as reading this *Revision Guide*, you might like to use some or all of the following strategies:
- Make your own condensed summary notes.
- Write key definitions onto flash cards.
- Work through facts until you can recall them.
- Ask your friends and family to test your recall.
- Make posters which cover items of the specification for your bedroom walls.
- Carry out exam practice.

Measure your revision in terms of the progress you are making rather than the length of time you have spent working. You will feel much more positive if you are able to say specific things you have achieved at the end of a day's revision rather than thinking 'I spent eight hours inside on a sunny day!' Don't sit for extended periods of time. Plan your day so that you have regular breaks, fresh air, and things to look forward to.

How to improve your recall

Here's a good strategy for recalling information:
- Focus on a small number of facts. Copy out the facts repeatedly in silence for five minutes then turn your piece of paper over and write them from memory.
- If you get any wrong then just write these out for another five minutes.
- Finally test your recall of all the facts.
- Come back to the same facts later in the day and test yourself again.
- Then revisit them the next day and again later in the week.

By carrying out this process the facts will become part of your long-term memory – you will have learnt them!

Once you have built up a solid factual knowledge base you need to test it by completing some past papers for practice. It might be a good idea to tackle several questions on the same topic from a number of papers rather than working through a whole paper at once. This will enable you to identify your weak areas so that you can work on them in more detail.

Finally, remember to complete some mock exam papers under exam conditions.

Answering exam papers

When you look at your exam paper read through all the questions. Identify which are the easiest for you to answer. Start by answering these questions.

Remember to read each question carefully and make sure you are answering the question that is actually set and not the one you would like to be set. Remember to look at the number of marks available for each question and tailor the number of points you make in your answer accordingly. *Do not* write an essay for a question that only attracts one or two marks!

With short-answer questions, look at the amount of space that has been left for the answer. This indicates the length of answer that the examiner anticipates you to give – depending on the size of your handwriting, of course.

Make every effort to answer all the questions. An unanswered question will always score 0! If you can, leave enough time to check through your answers at the end.

Here are some popular words which are often used in exam questions. Make sure you know what the examiner means when each of these words is used.

- **Describe** – Write down all the key points using words and, where appropriate, diagrams. Think about the number of marks that are available when you write your answer.

- **Calculate** – Write down the numerical answer to the question. Remember to include your working and the units.

- **State** – Write down the answer. Remember a short answer rather than a long explanation is required.

- **Suggest** – Use your biological knowledge to answer the question. This term is used when there is more than one possible answer or when the question involves an unfamiliar context.

- **Sketch** – When this word is used a simple freehand answer is acceptable. Remember to make sure that you include any important labels.

- **Define** – Write down what a biological term or statement means.

- **Explain** – Write down a supporting argument using your biological knowledge. Think about the number of marks that are available when you write your answer.

- **List** – Write down a number of points. Think about the number of points required.

- **Discuss** – Write down details of the points in the given topic.

A2 GCE Scheme of assessment

The table below shows how the marks are allocated in the OCR A2 Biology course. The A2 units account for 50% of the A2 marks.

Unit	Method of assessment	% of A2
A2 Unit F214: *Communication, Homeostasis and Energy*	1 hour 15 minutes written paper Candidates answer all questions. 60 marks	15%
A2 Unit F215: *Control, genomes and environment*	2 hour written paper Candidates answer all questions. 100 marks	25%
A2 Unit F216: *Practical Skills In Biology 2*	Coursework Candidates complete three tasks set by the exam board and marked internally. 40 marks	10%

Assessment objectives

Candidates are expected to demonstrate the following in the context of the biological content of A2 Biology:

Knowledge and understanding

- recognise, recall and show understanding of scientific knowledge;
- select, organise and communicate relevant information in a variety of forms.

Application of knowledge and understanding

- analyse and evaluate scientific knowledge and processes;
- apply scientific knowledge and processes to unfamiliar situations including those related to issues;
- assess the validity, reliability and credibility of scientific information.

How science works

- demonstrate and describe ethical, safe and skilful practical techniques and processes, selecting appropriate qualitative and quantitative methods;
- make, record and communicate reliable and valid observations and measurements with appropriate precision and accuracy;
- analyse, interpret, explain and evaluate the methodology, results and impact of their own and others' experimental and investigative activities in a variety of ways. investigative activities in a variety of ways.

1.01 Communication

Multicellular organisms have specialised organ systems, for example, for gaseous exchange, digestion, and transport. The activities of their various organs and systems need to be coordinated so that they work together and so that they can respond to changes both within and outside the organism. This coordination is brought about by specialised internal communication systems which take the form of nervous communication and chemical messages such as hormones. These are both examples of **cell signalling**.

Chemical signals in communication

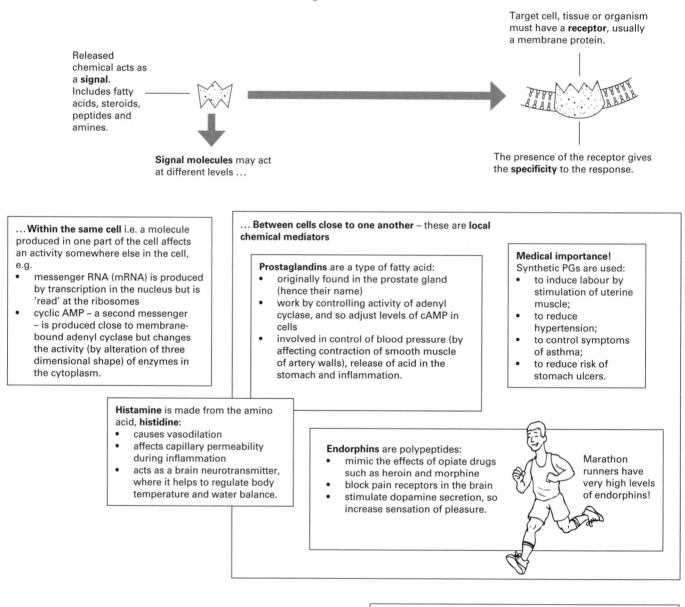

Released chemical acts as a **signal**. Includes fatty acids, steroids, peptides and amines.

Signal molecules may act at different levels …

Target cell, tissue or organism must have a **receptor**, usually a membrane protein.

The presence of the receptor gives the **specificity** to the response.

…Within the same cell i.e. a molecule produced in one part of the cell affects an activity somewhere else in the cell, e.g.
- messenger RNA (mRNA) is produced by transcription in the nucleus but is 'read' at the ribosomes
- cyclic AMP – a second messenger – is produced close to membrane-bound adenyl cyclase but changes the activity (by alteration of three dimensional shape) of enzymes in the cytoplasm.

… Between cells close to one another – these are **local chemical mediators**

Prostaglandins are a type of fatty acid:
- originally found in the prostate gland (hence their name)
- work by controlling activity of adenyl cyclase, and so adjust levels of cAMP in cells
- involved in control of blood pressure (by affecting contraction of smooth muscle of artery walls), release of acid in the stomach and inflammation.

Medical importance!
Synthetic PGs are used:
- to induce labour by stimulation of uterine muscle;
- to reduce hypertension;
- to control symptoms of asthma;
- to reduce risk of stomach ulcers.

Histamine is made from the amino acid, **histidine**:
- causes vasodilation
- affects capillary permeability during inflammation
- acts as a brain neurotransmitter, where it helps to regulate body temperature and water balance.

Endorphins are polypeptides:
- mimic the effects of opiate drugs such as heroin and morphine
- block pain receptors in the brain
- stimulate dopamine secretion, so increase sensation of pleasure.

Marathon runners have very high levels of endorphins!

… Between tissues: these are the typical HORMONES
- include steroids (e.g. oestrogens), amines (e.g. adrenaline) and polypeptides (e.g. insulin)
- control internal environment (homeostasis) e.g. insulin and blood glucose concentration
- control growth rate (e.g. human growth hormone)
- control sexual development and reproductive activity (e.g. oestrogen, testosterone, oxytocin)
- manage the body's response to stress (e.g. adrenaline, cortisone).

… Between different individuals of the same species: these are PHEROMONES
- many are fatty acids – they are volatile and disperse quickly into the environment
- can act over long distances (e.g. sex attractants) or very close to producer (e.g. queen substance in a hive)
- may be important in human sexual attraction, although we mask them with perfumes!

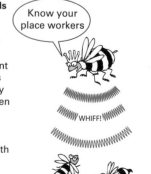

Know your place workers

WHIFF!

Electrical and chemical coordination

An individual's coordinated responses to changes in the environment (stimuli) are the result of the coordinated activities of the nervous system and endocrine system. Although their activities may be coordinated, the two systems work in different ways.

Nervous system	Endocrine system
• **Nerve impulses** are electrical and transmitted by nerve cells called **neurones**.	• **Hormones** are chemicals produced by different **endocrine glands** which secrete them into the bloodstream.
• Muscles or glands (called **effectors**) respond to nerve impulses.	• Hormones are transported in the bloodstream to all parts of the body. However, each hormone only affects its particular **target tissue** because only that tissue has receptors which bind to the hormone in question.
• Effectors respond to nerve impulses in milliseconds.	• The response of a target tissue to its particular hormone is long-lasting.

Nerves and the nervous system

Neurones are grouped into bundles called **nerves** which pass to all of the muscles and glands (the effectors) of the body. The nerves form the **nervous system**.

Question

1 The nervous system and endocrine system work in different ways. Summarize the differences.

An organism's *external* environment is its habitat. Its *internal* environment is the tissue fluid which bathes the cells of its body, and the blood which circulates through its blood vessels.

Keeping conditions (external and internal) constant is called **homeostasis**. Internally, conditions kept constant include

- core body temperature
- blood pH (acidity/alkalinity)
- blood glucose concentration

Recall that the term metabolism refers to all of the chemical reactions taking place in cells. *Recall* also that optimum temperature and optimum pH refer to the respective values at which enzymes are most active and the rates of enzyme catalysed reactions are therefore at a maximum. Keeping core body temperature and blood pH constant at optimum values which maximize enzyme activity means that the metabolism of cells is at its most efficient. Keeping blood glucose concentration constant at a value which optimizes energy transfer within cells also contributes to the efficiency of cell metabolism.

Recall that the water potential of a solution is a measure of the concentration of water making up that solution. *Recall* also that blood plasma and tissue fluid are solutions of different solutes and that the movement of water (osmosis) between them depends on the difference in value between their respective water potentials. This difference accounts for the movement of water and substances in solution between cells, tissue fluid and blood. Glucose is an important solute in blood plasma and tissue fluid. Keeping blood glucose concentration constant contributes to the controlled movement of water and substances in solution between tissues and their blood supply.

By maximizing the efficiency of cell metabolism and keeping the composition of tissue fluid and blood constant (within narrow limits), cells function efficiently.

> As a result the organism functions efficiently independently of changes (within wide limits) in its external environment.

> As a result the organism's chances of survival are improved.

For example, mammals and birds have a range of homeostatic mechanisms which enable them to live in extreme environments: polar bears and penguins survive sub-zero polar conditions; camels tolerate the blistering heat of deserts.

Homeostasis is the result of self regulating systems which work by means of **feedback mechanisms**. Refer to the diagram as you read about the characteristics of self-regulating systems:

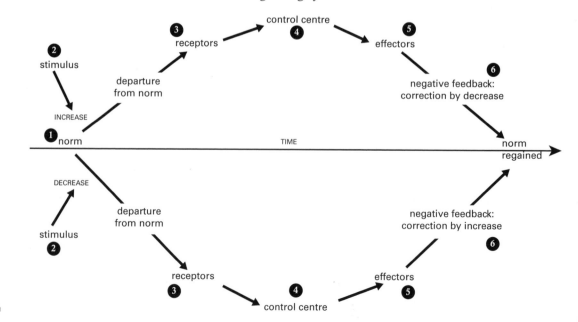

Characteristics of a self-regulating system

① **norm** (normal value) at which each system works. For example the blood glucose concentration of human blood is normally 90 mg glucose 100 cm⁻³ blood. Instead of norm, the terms **set point** or **reference point** are sometimes used.

② **stimuli** which cause deviations from the norm

③ **receptors** which detect deviations from the norm

④ **control centre** which receives information from receptors, coordinates the information, and sends instructions to effectors

⑤ **effectors** which bring about the responses needed to return the system to its norm

⑥ **feedback** which informs the receptors of the changes in the system caused by the effectors

Feedback ⑥ describes the situation where the information about changes in the system affects what happens to the changes in the future. When the information affects the system so that any change from the norm…

- reverses the direction of that change *towards* the norm, then we say that the feedback is **negative**
- causes more and more change *away* from the norm, then we say that the feedback is **positive**

Negative feedback maintains stability in a system. It controls the system so that conditions fluctuate around the norm. The system is self-adjusting. For example, human body temperature fluctuates and adjusts around a norm of 37°C. Homeostasis depends on the negative feedback mechanisms which enable systems to self-adjust.

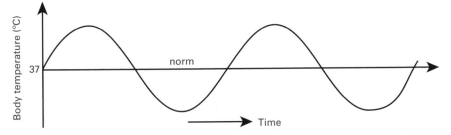

Positive feedback does *not* maintain stability in a system. It reinforces the original change and a chain reaction quickly develops. If positive feedback runs out of control then the system may destroy itself. In the real world, however, negative feedback of some sort eventually brings the self-reinforcing changes causing the chain reaction under control. For example, damaged tissues release substances that activate platelets in the blood. Activated platelets release substances which activate more platelets… and so on. A chain reaction quickly develops and a blood clot forms. This breaks the chain reaction and the norm is quickly re-established.

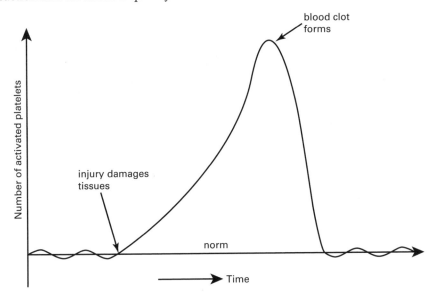

Fact file

Substances produced during certain bacterial infections temporarily shift the core body temperature above 37°C. The substances are called **pyrogens**. They affect the heat loss centre in the hypothalamus. The infected person appears flushed and produces a lot of sweat… both symptoms of a fever. If homeostatic control breaks down, then positive feedback reinforces further increase in core body temperature with possible fatal results.

Questions

1 What are the characteristics of self-regulating systems?

2 What is negative feedback?

3 Why does positive feedback not maintain stability in a system?

The heat that warms animal bodies comes from the

- Sun
- chemical reactions of metabolism

Ectotherms are animals which depend mainly on the sun as the source of body heat; **endotherms** depend mainly on their metabolism.

Birds and mammals are endotherms. High rates of metabolism release a lot of heat, which makes it possible to keep body temperature constant regardless of changes in the temperature of the environment. We call birds and mammals **homeotherms**.

The body temperature of other animals varies, often matching the temperature of the environment. We call them **poikilotherms**. The graph illustrates the point.

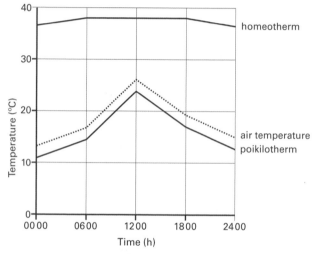

The body temperature of a homeotherm and a poikilotherm and the air temperature recorded over 24 hours

Fact file

Not all parts of the body of a homeotherm are kept at a constant temperature, only the **body's core**. The body's surface where heat is exchanged with the environment is always cooler than the core temperature. The term **thermoregulatory** refers to the processes which enable an animal to maintain a constant core body temperature regardless of changes in the temperature of the environment.

Some poikilotherms achieve control of core body temperature despite rates of metabolism lower than homeotherms. Control is often achieved through different behaviours. For example, reptiles such as snakes and lizards sun themselves on rocks.

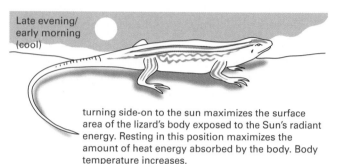

turning side-on to the sun maximizes the surface area of the lizard's body exposed to the Sun's radiant energy. Resting in this position maximizes the amount of heat energy absorbed by the body. Body temperature increases.

turning head-on to the Sun minimizes the surface area of the lizard's body exposed to the Sun's radiant energy. Resting in this position minimizes the amount of heat energy absorbed by the body. Body temperature is stabilized.

How lizards achieve some control of core body temperature

Producing and losing body heat

The diagram represents the energy inputs and outputs of animals that live on land. The Sun's heat and the heat produced through the metabolism of cells are the components of inputs. Outputs are the ways body heat is lost to the environment. An animal is able to maintain a constant body temperature when the heat it gains (inputs) is equal to the heat it loses (outputs).

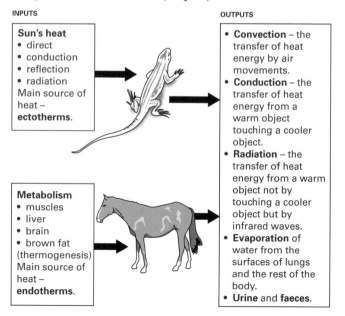

INPUTS

Sun's heat
- direct
- conduction
- reflection
- radiation
Main source of heat –
ectotherms.

Metabolism
- muscles
- liver
- brain
- brown fat (thermogenesis)
Main source of heat –
endotherms.

OUTPUTS
- **Convection** – the transfer of heat energy by air movements.
- **Conduction** – the transfer of heat energy from a warm object touching a cooler object.
- **Radiation** – the transfer of heat energy from a warm object not by touching a cooler object but by infrared waves.
- **Evaporation** of water from the surfaces of lungs and the rest of the body.
- **Urine** and **faeces**.

The faster the rate at which fats and sugars are oxidized in cells, the greater is the amount of heat energy released. Many of these oxidation reactions occur in cellular respiration and the rates of reaction are controlled by the hormone thyroxine produced and released by the thyroid gland. Cellular respiration is an important part of the cell's metabolism.

Thyroxine controls the metabolic rate of most endotherms at a level which enables them to maintain a constant core body temperature (usually 37°C in mammals; 41°C in birds). However, there is a cost: endotherms need much more food than ectotherms of equivalent size. For example, shrews (endotherms) eat food equivalent to their own body mass each day; cockroaches (ectotherms) can go without food for days. The extra food replaces the fats and sugars consumed in the metabolism which enables shrews to maintain a constant core body temperature.

Maintaining constant core body temperature

Remember that the temperature at the body's surface of an endotherm is cooler than the body's core.

The diagram summarizes the responses which enable endotherms (e.g. humans) to regulate their body temperature. The numbers ❸ to ❺ on the diagram and its checklist refer to the numbered characteristics of self-regulating systems listed on page 11.

In the lists:

- ❶ refers to the norm (normal value): in humans (and most mammals) the norm core body temperature is 37°C
- ❷ refers to the stimuli which cause deviations from the norm: here variations in the temperature of the environment cause changes in the temperature at the body's surface

Notice that the responses which regulate

- body surface (skin) temperature are voluntary responses
- core body temperature are involuntary responses

The homeostatic mechanisms controlling body surface temperature and core body temperature work together. Their interaction sets up voluntary and involuntary responses which compensate for changes in body surface temperature before any change in core temperature occurs.

Questions

1. Summarize the different ways body temperature is regulated in animals.
2. Explain the different ways an animal loses body heat to the environment.
3. How does the hormone thyroxine affect the release of heat by cells?

Checklist

Body surface temperature	Core body temperature
❸ Thermoreceptors in the...	
...skin detect changes in its temperature • heat receptors detect increases in temperature • cold receptors detect decreases in temperature	...hypothalamus of the brain detect changes in the temperature of the blood supplying the brain • heat loss centre is activated by increases in blood temperature • heat gain centre is activated by decreases in blood temperature
❹ The control centre in the...	
...cortex of the brain which enables us to think, feed and decide on our responses to stimuli (what we do). We say that the responses are voluntary.	...hypothalamus of the brain which not only detects changes in blood temperature but also sends signals via the autonomic nervous system controlling responses which do not require thinking and decision. We say that the responses are involuntary.
❺ The effectors which bring about the responses are...	
...the skeletal muscles which enable us to put our decisions into action ... putting on clothes or removing them depending on the skin temperature at which we feel comfortable	...the glands (sweat) and muscles (cause shivering, raise hairs, control blood flow) which enable the body to conserve or lose heat depending on its core temperature

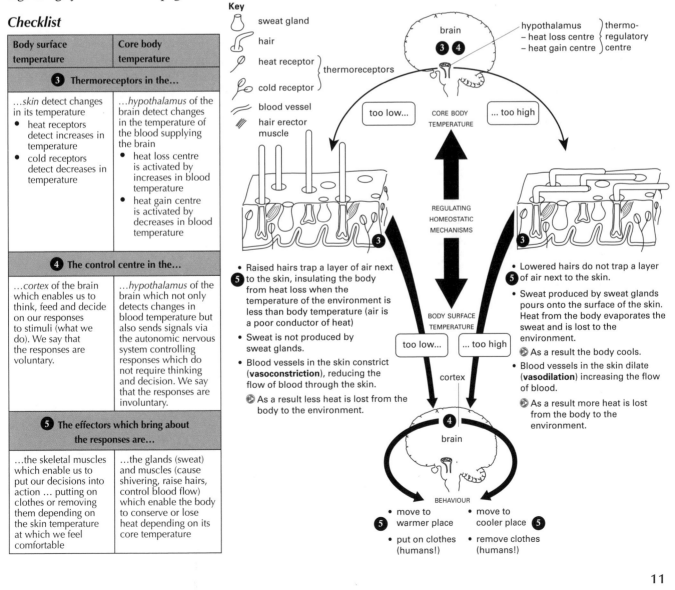

Key
- sweat gland
- hair
- heat receptor } thermoreceptors
- cold receptor
- blood vessel
- hair erector muscle

brain ❸ ❹

hypothalamus } thermo-
– heat loss centre } regulatory
– heat gain centre } centre

too low...   CORE BODY TEMPERATURE   ... too high

REGULATING HOMEOSTATIC MECHANISMS

BODY SURFACE TEMPERATURE

too low...   ... too high

cortex

- Raised hairs trap a layer of air next to the skin, insulating the body from heat loss when the temperature of the environment is less than body temperature (air is a poor conductor of heat)
- Sweat is not produced by sweat glands.
- Blood vessels in the skin constrict (**vasoconstriction**), reducing the flow of blood through the skin.
  - As a result less heat is lost from the body to the environment.

- Lowered hairs do not trap a layer of air next to the skin.
- Sweat produced by sweat glands pours onto the surface of the skin. Heat from the body evaporates the sweat and is lost to the environment.
  - As a result the body cools.
- Blood vessels in the skin dilate (**vasodilation**) increasing the flow of blood.
  - As a result more heat is lost from the body to the environment.

❹ brain

BEHAVIOUR

- ❺ move to warmer place
- ❺ move to cooler place
- put on clothes (humans!)
- remove clothes (humans!)

A **stimulus** is a change in the environment that causes an organism to take action. The action taken is the **response**. *Why do organisms respond to stimuli?* Improving chances of **survival** is the short answer. For example, by moving away, potential prey improve their chances of escaping predators; putting on more clothes prevents the body from becoming dangerously chilled.

The contraction of muscles pulling on a skeleton enables animals to respond to stimuli (prey moving away from a predator, for example). Plant responses are the result of growth movements. When the movements are the result of **directional stimuli** (stimuli coming mainly from one direction), then the response is called a **tropism**.

- Tropisms are **positive** if a plant grows towards the more intense source of a stimulus, and **negative** if it grows away. For example, shoots grow towards light where it is most intense (positive phototropism) and grow away from the force of gravity (negative geotropism). Roots grow to where water in the soil is most abundant (positive hydrotropism).

  - As a result the plant is more likely to survive: for example, the leaves of a shoot receive as much light as possible, maximizing the rate of photosynthesis; roots 'find' water where it is most abundant.

- The tropisms (positive or negative) are the result of differences in the growth rate of tissues on one side or the other of the shoots or roots in question.

The reflex arc

Nerve impulses transmitted by **neurones (nerve cells)** carry information about stimuli (detected by **sensory receptors**) to **effectors** (muscles and glands) which respond to the stimuli.

- **Sensory neurones** ending in sensory receptor cells transmit nerve impulses to the central nervous system.
- **Relay** (or **inter-**) **neurones** link sensory neurones with motor neurones. They are within the spinal cord and the brain.
- **Motor neurones** transmit nerve impulses from the central nervous system to effectors (muscles and glands) which respond by contracting (muscles) or secreting substances (glands).

Nerves consist of bundles of hundreds/thousands of neurones. Their simplest arrangement enables an individual to respond to a stimulus. The arrangement is called a **reflex arc**. The response is a **reflex response**. We say that it is **involuntary** because it is an automatic reaction to stimuli not under conscious control. For example, we automatically jerk our hand out of harm's way if we touch a hot stove.

In the diagram, each nerve of the reflex arc is represented by only one neurone; the sensory receptor by just one receptor cell. The numbers 1–5 track the sequence of events.

Questions

1 What is a stimulus? What is a response?

2 What is the difference between a positive tropism and a negative tropism?

3 Explain why a reflex response is said to be involuntary.

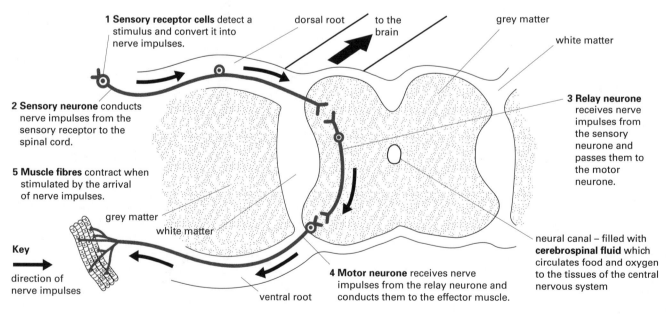

1 Sensory receptor cells detect a stimulus and convert it into nerve impulses.

dorsal root

to the brain

grey matter

white matter

2 Sensory neurone conducts nerve impulses from the sensory receptor to the spinal cord.

3 Relay neurone receives nerve impulses from the sensory neurone and passes them to the motor neurone.

5 Muscle fibres contract when stimulated by the arrival of nerve impulses.

grey matter

white matter

Key

→ direction of nerve impulses

neural canal – filled with **cerebrospinal fluid** which circulates food and oxygen to the tissues of the central nervous system

4 Motor neurone receives nerve impulses from the relay neurone and conducts them to the effector muscle.

ventral root

Section across the spinal cord: the neurones represent the arrangement of nerves which form a reflex arc

Sensory receptors

A sensory receptor is either the specialised ending of a single dendrite of a sensory neurone, or a separate cell that passes impulses onto a sensory neurone. It detects stimuli and forms the first link in the chain of events that brings about a response. The cells of sensory receptors are biological **transducers**, converting one form of energy (the stimulus) into electrical energy (action potentials) to which the body can respond. Animal sensory receptors are categorized according to the stimuli they detect.

- **Mechanoreceptors** detect physical force such as pressure and stretch (touch, muscle contraction/relaxation).
- **Photoreceptors** detect light (vision).
- **Thermoreceptors** detect changes in temperature.
- **Chemoreceptors** detect substances in solution (taste/smell).

Pacinian corpuscles

Lying deep within the skin, **Pacinian corpuscles** detect pressure when the skin is firmly touched. They are mechanoreceptors. Each contains the single end of the nerve fibre of a dendrite of a sensory neurone, wrapped around by layers of membrane called **lamellae**. A jelly-like material separates the layers.

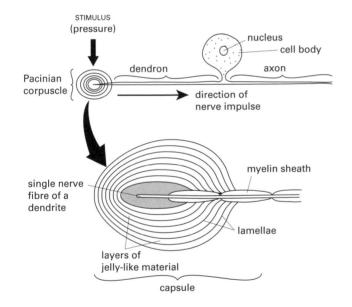

A sensory neurone and its Pacinian corpuscle. The sensory dendrite is enclosed by layers of lamellae forming a capsule.

The pressure of a firm touch (stimulus) deforms (stretches and changes the shape of) the capsule of a Pacinian corpuscle. The pressure must be enough to deform the dendrite as well as the capsule.

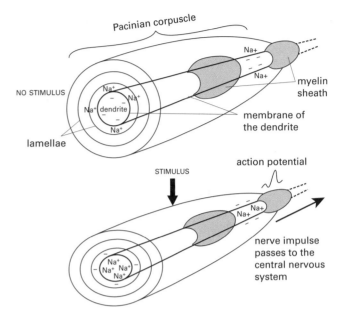

- At rest the concentration of sodium ions (Na^+) on the outer surface of the membrane of the dendrite is greater than the inner surface.
- Deformation of a localized part of the membrane temporarily alters its permeability to Na^+.
  - As a result Na^+ diffuses down its concentration gradient from the outer surface to the inner surface of the membrane.
  - As a result the inner surface becomes less negatively charged. We say that the membrane is **depolarized**.
- Localized depolarization in a Pacinian corpuscle (and other sensory receptor cells) is called a **generator potential**.
- The more intense (stronger) the stimulus, the greater is the generator potential.
- When the generator potential reaches (or is more than) a threshold value, it triggers an action potential in the sensory neurone attached to the Pacinian corpuscle.

Rods and cones

Cells called **rods** and **cones** line the retina of the human eye. The cells are photoreceptors. They convert light energy into the electrical energy of generator potentials. Rods and cones are connected to a network of other neurones in the retina. These neurones form the fibres of the optic nerve, which passes from each eye to the visual cortex of the brain. The generator potentials formed in the rods and cones trigger action potentials in each optic nerve. The action potentials are transmitted as nerve impulses along each optic nerve to the visual cortex. Here they are interpreted as images of the object we are looking at.

The neurone

Recall that each nerve of the nervous system consists of a bundle of neurones. The diagram shows a human motor neurone.

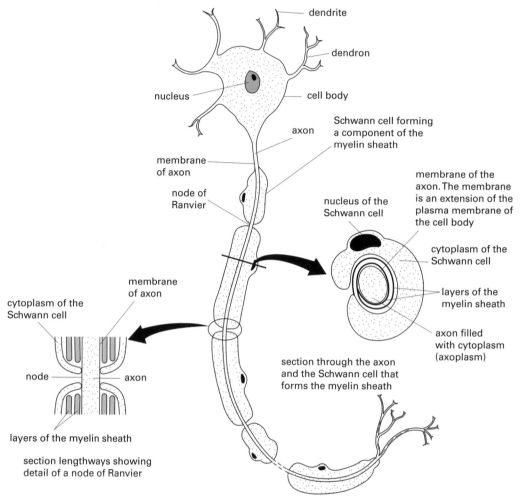

- A **dendron** and its **dendrites** are thin extensions of the **cell body** which carry nerve impulses *towards* the cell body.
- **Axon** – an extension of the cell body that carries nerve impulses *away* from the cell body.
- **Schwann cell** – a type of cell that forms the components of the myelin sheath during the development of the nervous system.
- **Myelin sheath** – forms from Schwann cells. It consists of layers of membrane wrapped round the axon. Myelin is a fatty substance and an important component of the sheath. It insulates the axon.
- **Node of Ranvier** – a break in the myelin sheath where the axon is uncovered. Nerve impulses jump from node to node, speeding up their passage along the axon.

Resting potential

Remember that the axon is an extension of the cell body of a neurone. It is surrounded by a membrane which is an extension of the plasma membrane surrounding the cell body.

The diagram shows the distribution of ions on either side of the membrane of the axon of a neurone at rest (not stimulated). The concentration of

- sodium ions (Na^+) is higher on the outer surface of the membrane than on the inner surface
- potassium ions (K^+) is higher on the inner surface of the membrane than on its outer surface

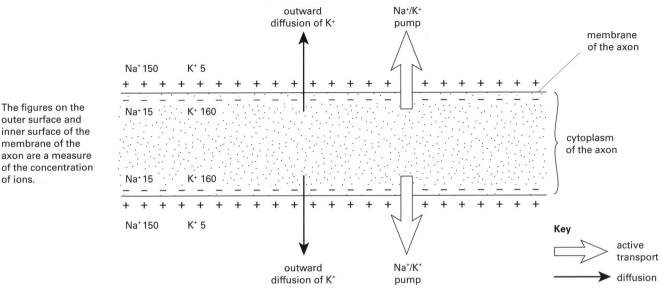

The figures on the outer surface and inner surface of the membrane of the axon are a measure of the concentration of ions.

The distribution of ions on both sides of the membrane of an axon at rest

Since the concentration of potassium ions is much greater on the inner surface of the membrane compared with the outer surface, potassium ions rapidly diffuse out of the axon. Remember that potassium ions are positively charged, (K⁺). The outward diffusion of positive ions, therefore, means that the inner surface of the membrane becomes slightly negative relative to its outer surface. As more potassium ions pass to the outer surface of the membrane, their diffusion outwards slows until equilibrium is reached. At this point the rate of diffusion of potassium ions out of and into the axon across its membrane is balanced.

How is the difference in concentration of the different ions across the membrane maintained? Active transport makes the difference.

- Sodium ions and potassium ions are actively exchanged by the **sodium–potassium pump** located in the membrane of the axon. For every three sodium ions removed from the axon by the sodium–potassium pump, two potassium ions are brought in.
    - As a result there are more positive ions on the outer surface of the membrane than on its inner surface.
    - As a result the negativity of the inner surface of the membrane established by the outward diffusion of potassium ions is maintained.

The difference (electrical potential of the inner surface of the membrane of the axon – the electrical potential of its outer surface) is called the **potential difference** and is usually about –65 mV. The value represents the **resting potential** of the membrane and overall the axon is said to be **polarized**.

Fact file

A resting potential of –65 mV means that the electrical potential of the inner surface of the membrane of the axon is 65 mV lower than the outer surface. Accumulation of negatively charged organic ions (mostly of the amino acids aspartate and glutamate) along the inner surface of the membrane helps to maintain its negativity. The ions do not diffuse out of the axon because its membrane is impermeable to them.

Qs and As

Q Why do many more potassium ions diffuse out of the axon than sodium ions diffuse in?

A *The membrane of the axon is much more permeable to potassium ions than sodium ions.*

Questions

1 What is the relationship between Schwann cells and the myelin sheath wrapped round the axon of a neurone?

2 Explain how the resting potential of an axon is maintained.

3 What is the difference between a dendron and an axon?

The diagram shows that the potential difference across the membrane of the axon of a stimulated neurone changes. The change in potential difference reverses the resting potential and is called an **action potential**.

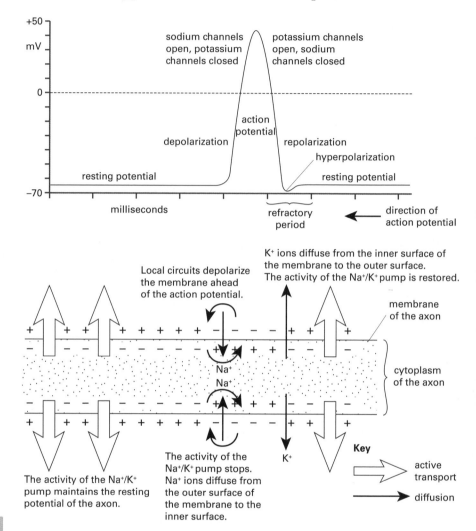

An action potential: the trace shows resting and action potentials recorded by a cathode ray oscilloscope (the instrument is used to measure small electrical changes). The changes in the movement of ions across the membrane of the axon correspond to the trace. Notice that the nerve impulse is moving from right to left.

Fact file

The channels in the membrane which allow the passage of sodium ions and potassium ions are closed when the membrane is in its resting state (polarized). When the membrane is depolarized they open, allowing the passage of ions. The channels are said to be **voltage gated** because they open only when the potential difference of the membrane of the axon changes following a stimulus.

The fact that the sodium channels open more quickly than the potassium channels accounts for the

- *depolarization* of the membrane when sodium ions enter the axon

- *repolarization* of the membrane when potassium ions leave the axon

- The action of the sodium–potassium pump in the membrane of the axon stops.

  - As a result the permeability of the membrane to sodium ions and potassium ions changes. Ion channels open in the membrane and the ions diffuse along their respective concentration gradients produced when the axon was at rest.

- Sodium channels are the first to open. Sodium ions diffuse from where they are in greater concentration on the outer surface of the membrane to the inner surface where they are in lower concentration.

  - As a result the inner surface of the membrane becomes less negative. We say that the membrane is **depolarized**.

- Positive charge builds up on the inner surface of the membrane as sodium ions continue to diffuse into the axon.

- When the electrical potential of the inner surface of the membrane reaches +40 mV compared with the outer surface, the sodium channels close.

  - As a result sodium ions stop diffusing into the axon.

- Potassium channels in the membrane then open. Potassium ions diffuse from where they are in greater concentration on the inner surface of the membrane to the outer surface where they are in lower concentration.

  - As a result the inner surface of the membrane becomes more negative. We say that the membrane is **repolarized**.

- Potassium ions continue to diffuse from the inner surface of the membrane to its outer surface. This diffusion makes the potential difference across the membrane even more negative than the resting potential of –65 mV (**hyperpolarization**). Checking the diagram shows you what happens.
- Activity of the sodium–potassium pump in the axon membrane is restored.
    - As a result the distribution of sodium ions and potassium ions along the outer and inner surface of the membrane is restored.
    - As a result the potential difference across the membrane is restored to its resting state of –65 mV.

Recovery

The removal of potassium ions from the axon marks the beginning of the recovery phase. The membrane of the axon behind the action potential is repolarized and its resting potential restored. Reactivation of the sodium–potassium pump completes the process.

The period of recovery is called the **refractory period**. It lasts for about 1 ms, during which the generation of new action potentials is not possible. As a result the refractory period determines

- the **frequency** (number per unit time) with which action potentials are transmitted along the axon
- the direction of the action potentials, i.e. *from* the cell body to the end of its axon. Action potentials cannot travel in the opposite direction because the membrane behind the action potential is refractory during the restoration of its resting potential.

Transmission of action potentials

Notice in the diagram the leading edge of the depolarized region of the membrane. Here **local circuits** cause depolarization of the resting part of the membrane ahead of the action potential. In this way the action potential moves (is **conducted**) along the axon. The conduction of an action potential is the **nerve impulse**.

The speed of conduction depends on the

- *myelin sheath* – its presence speeds up conduction
- *diameter of the axon* – the greater it is, the greater is the speed of conduction.

Axons covered with myelin are said to be **myelinated**. The fatty component of myelin electrically insulates the axon. *Remember* that the myelin sheath of a neurone is constricted at intervals along its length. Each constriction is called a **node of Ranvier**. The myelin sheath is broken and the axon uncovered.

Depolarization and the formation of action potentials cannot take place where the axon is myelinated. However action potentials can form at the nodes of Ranvier. Action potentials therefore jump from node to node (called **saltatory conduction**) speeding up their conduction. The speed of conduction reaches $120\,m\,s^{-1}$ for some myelinated axons compared with $0.5\,m\,s^{-1}$ for non-myelinated ones.

Action potentials only form if the stimulus is strong enough to begin depolarization of the membrane of the axon. In other words the stimulus must have a **threshold** value. A sub-threshold stimulus on its own is ineffective but a series quickly repeated may have a cumulative effect sufficient to initiate an action potential. The process is called **summation**.

Once started, the amplitude of the action potential remains the same at +40 mV as it travels along an axon, regardless of the strength of the stimulus above its threshold value. In other words, the formation of an action potential is an **all-or-nothing** response. It is the frequency of action potentials rather than their size that gives information about the intensity of the stimulus.

Qs and As

Q Why is the generation of new action potentials not possible during the refractory period?

A *Hyperpolarization at the beginning of the refractory period makes the generation of new action potentials impossible no matter how intense the stimulus.*

Why? The sodium channels are closed and depolarization cannot take place.

Qs and As

Q Why is an action potential an all-or-nothing response regardless of the strength of a stimulus above its threshold value?

A *The depolarization and repolarization of the membrane of an axon depends on the concentration gradients of sodium ions and potassium ions, and the timing of the openings of the voltage-gated ion channels. The concentration gradients and timings are fixed so the size of the action potential is constant, regardless of the strength of a stimulus above its threshold value.*

Questions

1 How does an axon become depolarized?
2 What is hyperpolarization?
3 Briefly explain the outcomes of the refractory period on the transmission of nerve impulses.

Propagation of action potentials in non-myelinated and myelinated neurones

Non-myelinated neurones

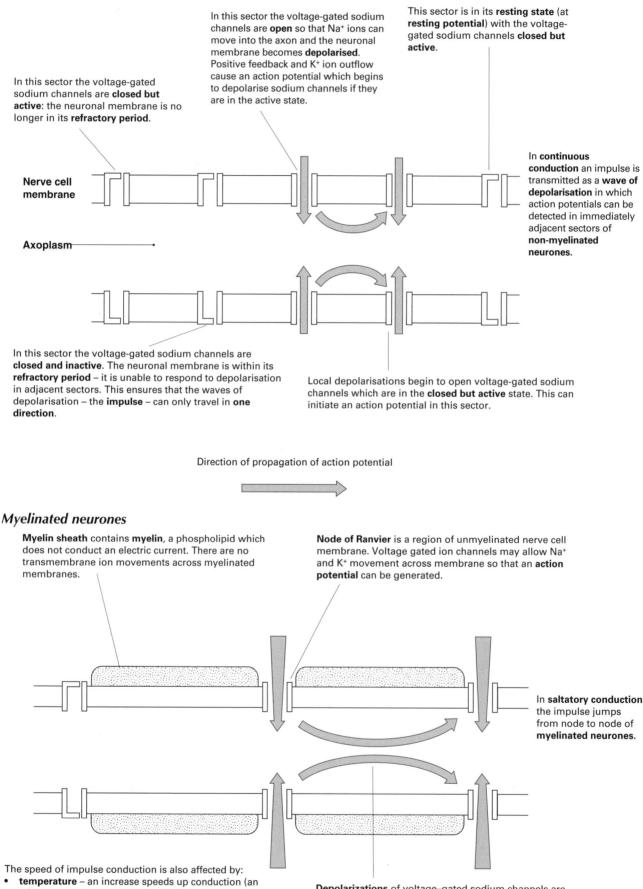

In this sector the voltage-gated sodium channels are **open** so that Na+ ions can move into the axon and the neuronal membrane becomes **depolarised**. Positive feedback and K+ ion outflow cause an action potential which begins to depolarise sodium channels if they are in the active state.

This sector is in its **resting state** (at **resting potential**) with the voltage-gated sodium channels **closed but active.**

In this sector the voltage-gated sodium channels are **closed but active**: the neuronal membrane is no longer in its **refractory period**.

Nerve cell membrane

Axoplasm

In **continuous conduction** an impulse is transmitted as a **wave of depolarisation** in which action potentials can be detected in immediately adjacent sectors of **non-myelinated neurones**.

In this sector the voltage-gated sodium channels are **closed and inactive**. The neuronal membrane is within its **refractory period** – it is unable to respond to depolarisation in adjacent sectors. This ensures that the waves of depolarisation – the **impulse** – can only travel in **one direction**.

Local depolarisations begin to open voltage-gated sodium channels which are in the **closed but active** state. This can initiate an action potential in this sector.

Direction of propagation of action potential

Myelinated neurones

Myelin sheath contains **myelin**, a phospholipid which does not conduct an electric current. There are no transmembrane ion movements across myelinated membranes.

Node of Ranvier is a region of unmyelinated nerve cell membrane. Voltage gated ion channels may allow Na+ and K+ movement across membrane so that an **action potential** can be generated.

In **saltatory conduction** the impulse jumps from node to node of **myelinated neurones**.

The speed of impulse conduction is also affected by:
* **temperature** – an increase speeds up conduction (an advantage in being an endotherm);
* **diameter of axon** – an increase speeds up conduction (common in invertebrates, including the giant squid).

Depolarizations of voltage–gated sodium channels are effective over greater distances i.e. between adjacent nodes of Ranvier.

The synapse

The axon of a neurone ends in swellings called **synaptic knobs**. A minute gap called the **synapse** separates the knobs from the dendrites of the next neurone in line. Each knob contains numerous mitochondria and structures called **synaptic vesicles**. The membrane round a synaptic knob is called the **pre-synaptic** membrane. The membrane of a dendrite of the next neurone in line is called the **post-synaptic** membrane. In between each membrane is the narrow gap of the **synaptic cleft**. The gap is about 10 nm wide. Information must pass across each synapse from one neurone to the next for effectors to be able to respond to stimuli. The transmission of information across most synapses is chemical.

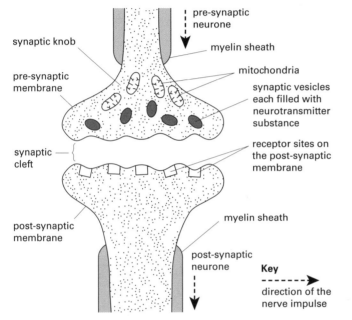

The structure of a synapse – acetylcholine molecules bind with the receptor sites on the post-synaptic membrane because the shape of molecule and receptor match

The diagram illustrates the structure of a chemical synapse. The synaptic vesicles each contain the chemical which diffuses across the synapse. The chemical is called **neurotransmitter** substance. **Acetylcholine** is an example of neurotransmitter substance. Neurones whose synapses depend on acetylcholine are called **cholinergic** neurones.

Transmission across the synapse

- The arrival of an action potential at a synaptic knob opens calcium channels in its pre-synaptic membrane, making it more permeable to calcium ions (Ca^{2+}) present in the synaptic cleft. We say that the channels are **voltage gated**.
- The calcium ions rapidly diffuse into the synaptic knob. They cause the synaptic vesicles filled with molecules of acetylcholine to move to the pre-synaptic membrane and fuse with it. The vesicles empty (an example of exocytosis) acetylcholine into the synaptic cleft.
- The molecules of acetylcholine diffuse across the synaptic cleft and bind to their specific receptors on the post-synaptic membrane.
- The binding of acetylcholine molecules to their receptors opens up sodium ion channels in the post-synaptic membrane, allowing the influx of sodium ions. Depolarization of the membrane occurs.
- The resting potential of the post-synaptic membrane is reversed and a new potential is generated.
- The new potential is called the **excitatory post-synaptic potential** (**EPSP**).

Fact file

The numerous mitochondria in the pre-synaptic knob generate the ATP required for the active transport of ions and refilling of synaptic vesicles with neurotransmitter substance. The presence of synaptic vesicles *only* in the synaptic knob and the location of receptors to which neurotransmitter substance binds *only* on the post-synaptic membrane means that synaptic transmission occurs *only* one way: pre-synaptic neurone → post-synaptic neurone.

On its own an EPSP normally does not produce sufficient depolarization to reach the threshold required to generate an action potential in the post-synaptic neurone. A number of EPSPs are required. EPSPs build up as more and more neurotransmitter substance binds to the receptors on the post-synaptic membrane. Threshold is reached when sufficient depolarization occurs in the post-synaptic membrane to generate an action potential in the neurone as a whole. The additive effect of EPSPs is called **temporal summation**.

So far the description of events refers to **excitatory synapses** but other synapses work in a different way. For example, **inhibitory synapses** respond to neurotransmitter not by promoting the inflow of sodium ions into the post-synaptic neurone and the depolarization of its post-synaptic membrane, but by promoting the outflow of potassium ions causing the polarization of the post-synaptic membrane to increase (**hyperpolarization**). Its threshold value is therefore greater, making the generation of an action potential in the post-synaptic neurone less likely. A post-synaptic neurone may be served by both types of synapse. Its activity is the result of the sum of their different inputs.

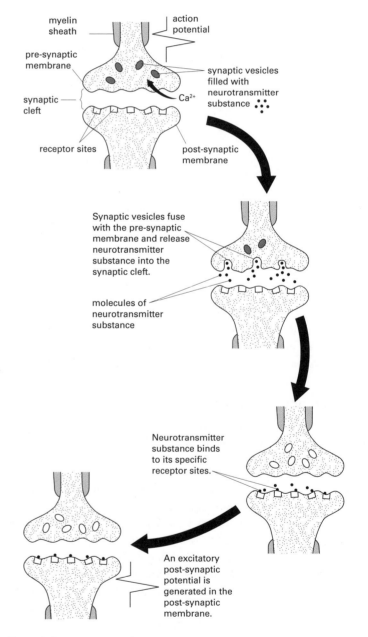

myelin sheath

action potential

pre-synaptic membrane

synaptic vesicles filled with neurotransmitter substance

synaptic cleft

Ca²⁺

receptor sites

post-synaptic membrane

Synaptic vesicles fuse with the pre-synaptic membrane and release neurotransmitter substance into the synaptic cleft.

molecules of neurotransmitter substance

Neurotransmitter substance binds to its specific receptor sites.

An excitatory post-synaptic potential is generated in the post-synaptic membrane.

Transmission across an excitatory synapse

After synaptic transmission

If neurotransmitter were not removed from the synaptic cleft, its presence would stimulate the repeated generation of action potentials in the post-synaptic neurone and the continuous stimulation of the effector at the end of the chain of events.

If the effector is a muscle, it would be permanently contracted: a condition known as **tetanus**. Breaking down neurotransmitter substance prevents such an outcome.

- The enzyme **cholinesterase** located on the post-synaptic membrane catalyses the breakdown (hydrolysis) of acetylcholine into ethanoic (acetic) acid and choline.
- These products of hydrolysis pass back into the pre-synaptic knob.
- Neurotransmitter substance is re-synthesized in the pre-synaptic knob from the products and incorporated into synaptic vesicles.
- Calcium ions (Ca^{2+}) are actively transported out of the pre-synaptic knob, re-establishing their concentration gradient across the pre-synaptic membrane.

The effect of drugs on the synapse

Different drugs have their effect on the receptors of the post-synaptic membrane. Their action increases or reduces synaptic transmission.

- **Agonists** are drugs that mimic the structure of molecules of neurotransmitter substance and combine with and activate receptors. Their effect promotes the generation of action potentials. They act as stimulants.
- **Antagonists** are drugs which also mimic neurotransmitters but when they bind to receptors do not activate them. They block the generation of action potentials. They act as tranquillizers.

The roles of synapses in the nervous system

Synapses transmit action potentials from one nerve cell to the next, allowing rapid communication in complex interconnected networks of nerve cells. They also have a vital role of coordination and control within the complexity of the nervous system. All the nerve cells of the body are interconnected by both inhibitory and excitatory synapses whose function is affected by fatigue and also by drugs. Synapses are also centrally involved in the processes of memory and learning within the brain.

Questions

1 Summarize the structure of a synapse.
2 What is an excitatory post-synaptic potential (EPSP)?
3 What is the difference between an excitatory synapse and an inhibitory synapse?

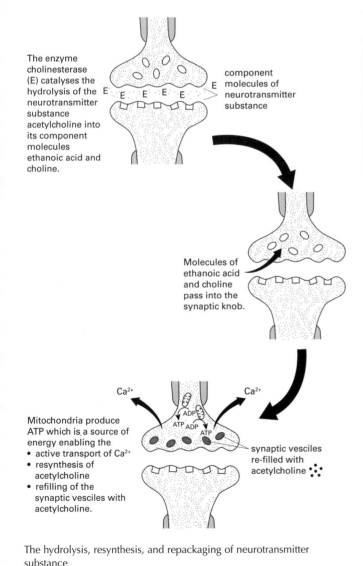

The enzyme cholinesterase (E) catalyses the hydrolysis of the neurotransmitter substance acetylcholine into its component molecules ethanoic acid and choline.

component molecules of neurotransmitter substance

Molecules of ethanoic acid and choline pass into the synaptic knob.

Mitochondria produce ATP which is a source of energy enabling the
- active transport of Ca^{2+}
- resynthesis of acetylcholine
- refilling of the synaptic vesicles with acetylcholine.

synaptic vesicles re-filled with acetylcholine

The hydrolysis, resynthesis, and repackaging of neurotransmitter substance

1.08 Hormones

Hormones and endocrine glands

The blood system is the link between a hormone and its target tissue. The sequence reads:

$$\text{endocrine gland} \xrightarrow{\text{produces}} \text{hormone} \xrightarrow{\text{circulates}} \text{blood} \xrightarrow{\text{affects}} \text{target tissue}$$

Hormones are chemical messengers synthesized in **endocrine glands** and secreted directly into the capillary vessels which supply blood to each gland. This is why endocrine glands are called **ductless glands**. In contrast, in **exocrine glands** the hormone is secreted through a duct, or tube, to its target tissue. The diagram shows where endocrine glands are located in the body.

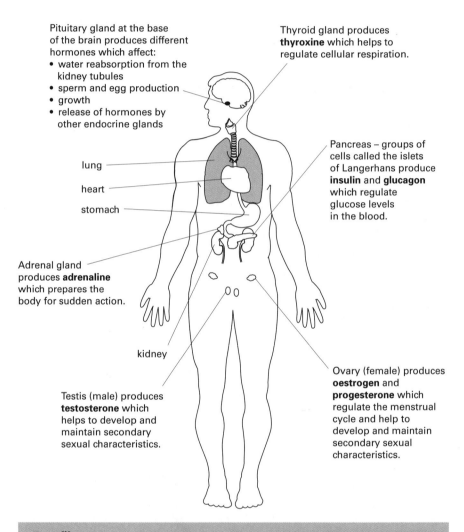

Pituitary gland at the base of the brain produces different hormones which affect:
• water reabsorption from the kidney tubules
• sperm and egg production
• growth
• release of hormones by other endocrine glands

Thyroid gland produces **thyroxine** which helps to regulate cellular respiration.

lung

heart

stomach

Pancreas – groups of cells called the islets of Langerhans produce **insulin** and **glucagon** which regulate glucose levels in the blood.

Adrenal gland produces **adrenaline** which prepares the body for sudden action.

kidney

Testis (male) produces **testosterone** which helps to develop and maintain secondary sexual characteristics.

Ovary (female) produces **oestrogen** and **progesterone** which regulate the menstrual cycle and help to develop and maintain secondary sexual characteristics.

Fact file

A gland consists of a cluster of cells which produces and secretes (releases) one or more substances with widespread effects. The substance(s) produced is the secretion. The action of most hormones is long-lasting. The exception is adrenaline which rapidly prepares the body for sudden action.

Questions

1 The nervous system and endocrine system work in different ways. Summarize the differences.

2 How are local chemical mediators different from hormones?

3 Why does aspirin reduce inflammation?

Local chemical mediators

Like hormones, **local chemical mediators** carry information. Unlike hormones they are not transported in the blood to target tissues elsewhere in the body, but affect the cells that synthesize them or cells nearby. **Histamine** and **prostaglandins** are examples of local chemical mediators.

- Histamine is released in response to the localized presence of an antigen (a substance which the body does not recognize as its own): the poison of a stinging nettle for example. It is produced by **white blood cells** and **mast cells** which are found in connective tissue. Its effects include pain, heat, redness, and inflammation. We say that the effect is a **non-specific immune response** because histamine is released in response to many different antigens, not just a particular antigen.
- There are about 30 different prostaglandins synthesized from fatty acids by most types of cells throughout the body. Their functions are various and often with opposite effects depending on the types of prostaglandin. For example, depending on type, some promote or inhibit
  - the body's inflammatory responses
  - blood clotting
  - contraction of the muscles of the wall of the intestine

Other prostaglandins induce (bring on) labour, which begins the birth process, relax the muscles in the walls of blood vessels reducing blood pressure, and increase blood flow in the kidneys.

Ibuprofen and aspirin reduce inflammation. *Why?* The drugs inhibit the secretion of prostaglandins.

Adrenaline and the adrenal glands

The medulla of the adrenal glands secretes adrenaline, the 'fight or flight' hormone which prepares the body for action. It stimulates:

- the liver to convert more glycogen to glucose
- deeper, more rapid breathing
- faster heartbeat
- diversion of blood from the gut to muscles.

These actions all help provide more glucose and oxygen for the working muscles. It may also cause

- dilation of the pupils
- raising of the hair.

Adrenaline does not enter its target cells but acts through a **second messenger**, cyclic AMP.

> ### Functions of the adrenal glands
>
> The adrenal cortex secretes adrenaline. This is not the only hormone produced by the adrenal glands. The adrenal medulla secretes steroid hormones:
>
> - aldosterone, which regulates the levels of sodium and potassium ions in the blood
> - cortisol, which suppresses the immune response to combat stress
> - androgens, which promote the development of the male secondary sexual characteristics.

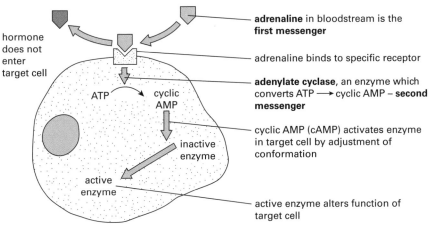

adrenaline in bloodstream is the **first messenger**

hormone does not enter target cell

adrenaline binds to specific receptor

adenylate cyclase, an enzyme which converts ATP ⟶ cyclic AMP – **second messenger**

cyclic AMP (cAMP) activates enzyme in target cell by adjustment of conformation

active enzyme alters function of target cell

Adrenaline works through a second messenger

Normally the concentration of glucose in the blood is about 90 mg glucose 100 cm^{-3} blood. However the value fluctuates according to circumstances.

- Following a meal the concentration of blood glucose increases as digested food is absorbed from the intestine into the bloodstream.
- During exercise the concentration of blood glucose decreases as glucose passes from the blood to vigorously contracting muscles, where it is used in cellular respiration.

Evening out fluctuations in blood glucose concentration is an example of homeostasis. The liver, muscles, and different hormones play an important role in the process.

Glucose in the body is stored as glycogen in liver and muscle. Different enzymes catalyse its

- synthesis: glucose → glycogen
- breakdown: glycogen → glucose

Different hormones control the activity of the enzymes.

Hormone	Target tissue	Source	Action
insulin	liver and muscle	pancreas	glucose → glycogen
glucagon	liver	pancreas	glycogen → glucose
adrenaline	muscle	adrenal glands	glycogen → glucose

The role of insulin and glucagon

The **pancreas** lies below the stomach in the first fold of the duodenum. It is an exocrine gland producing different digestive enzymes, which pass through the pancreatic duct to the duodenum. It is also an endocrine gland containing clusters of cells called the **islets of Langerhans**, which secrete hormones directly into the bloodstream.

Insulin is secreted by the numerous small **beta (ß) cells** of the islets in response to a *high* concentration of blood glucose (hyperglycaemia).

- Insulin makes the plasma membrane of liver cells and muscle cells more permeable to glucose.

  As a result more glucose is taken up from the blood by the cells.

- It activates **glycogen synthase** (and several other enzymes), which catalyses the condensation of glucose molecules forming glycogen in liver cells and muscle cells: glucose → glycogen.

The process is called **glycogenesis**.

- It promotes the conversion of glucose into lipids.

Overall, insulin *reduces* blood glucose concentration.

Glucagon is secreted by the less numerous, larger **alpha (α) cells** of the islets in response to a *low* concentration of blood glucose (hypoglycaemia).

- Glucagon reduces the permeability of the plasma membrane of liver cells to glucose.

  As a result less glucose is taken up from the blood by the cells.

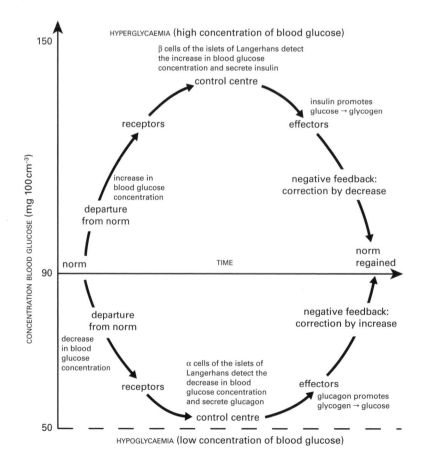

- Glucagon inhibits glycogen synthase (and other enzymes) catalysing the reactions of glycogenesis.
- It activates the enzyme **glycogen phosphorylase**, which catalyses the breakdown of glycogen in liver cells forming glucose: glycogen → glucose.

The process is called **glycogenolysis**.

- It activates **fructose bisphosphate phosphatase** (and other enzymes), which catalyses reactions that convert non-carbohydrate substances into glucose. The process is called **gluconeogenesis**.

Overall, glucagon *increases* blood glucose concentration.

Adrenaline secreted by the adrenal glands also affects the concentration of blood glucose. In response to a low concentration of blood glucose it

- *inactivates* glycogen synthase
- *activates* glycogen phosphorylase

  🜨 As a result blood glucose concentration *increases*.

Messengers and receptors

Remember:

- Hormones are **messenger molecules**. Insulin, glucagon, and adrenaline are examples.
- Receptors are proteins to which messenger molecules bind.

The different receptors which bind molecules of insulin, glucagon, and adrenaline are embedded in the phospholid bilayer of the plasma membrane of the cells of the hormones' target tissues.

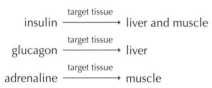

$$insulin \xrightarrow{\text{target tissue}} \text{liver and muscle}$$
$$glucagon \xrightarrow{\text{target tissue}} \text{liver}$$
$$adrenaline \xrightarrow{\text{target tissue}} \text{muscle}$$

When insulin binds with its receptor

- endocytosis carries the hormone-receptor complex into the cytoplasm of the target cell
- the complex stimulates the Golgi apparatus to bud off portions of material containing glucose carrier proteins
- the material passes to the cell surface where it fuses with the plasma membrane, increasing the number of glucose carrier proteins.

  🜨 As a result the uptake of glucose by the cell increases.

When glucagon and adrenaline combine with their respective receptors

- the enzyme **adenylate cyclase** is activated (adenylate cyclase is linked to the receptor protein)
- the enzyme catalyses the conversion of ATP to **cyclic (c)AMP**, which is an example of a **second** messenger molecule (hormone molecules are the **first** messenger)
- cAMP activates the enzyme glycogen phosphorylase.

  🜨 As a result the enzyme catalyses the breakdown
  $$glycogen \rightarrow glucose$$
  🜨 As a result glucose is released from the liver into the bloodstream.
  🜨 As a result blood glucose concentration increases.

Diabetes

Without treatment, a person with **diabetes** may suffer from the different symptoms of hyperglycaemia.

- The pH of the blood falls, causing **acidosis**.
- The volume of water lost in the urine increases and is excessive.
- The blood supply to the body's extremities is reduced.

In the long term, **gangrene** may develop in fingers and toes deprived of the oxygen and nutrients carried in the blood.

There are two forms of diabetes:

- **Type 1 or insulin-dependent diabetes** (sometimes called **juvenile-onset diabetes**) where the pancreas does not produce enough insulin. The deficiency may be the result of
  ○ the individual's immune system destroying the β cells of the pancreas – an example of an auto-immune disease, or …
  ○ the gene encoding the synthesis of insulin is faulty – an example of a genetic disorder.
- **Type 2 or non-insulin-dependent diabetes** (sometimes called **late-onset diabetes**) where the pancreas produces enough insulin, at least to begin with, but the body's tissues become insensitive to it.
  🜨 As a result tissues cannot make use of blood glucose as a source of energy.
  🜨 As a result the β cells of the islets produce more insulin and the liver releases more glucose.
- Eventually the β cells become less able to produce enough insulin and tissues become more resistant to it.
  🜨 As a result the blood glucose concentration increases.

Treatment depends on the form of diabetes. Daily injections of insulin together with a healthy balanced diet regulate the blood glucose levels of individuals with Type 1 insulin-dependent diabetes. For those with Type 2 non-insulin-dependent diabetes successful control is often possible through a healthy balanced diet alone.

Questions

1 What is the difference between glycogenesis, glycogenolysis, and gluconeogenesis?
2 What happens when a molecule of insulin binds to its receptor protein embedded in the plasma membrane of a liver cell?
3 Why is cyclic (c) AMP called a secondary messenger molecule? What does cAMP do?

The pancreas: an endocrine and exocrine gland

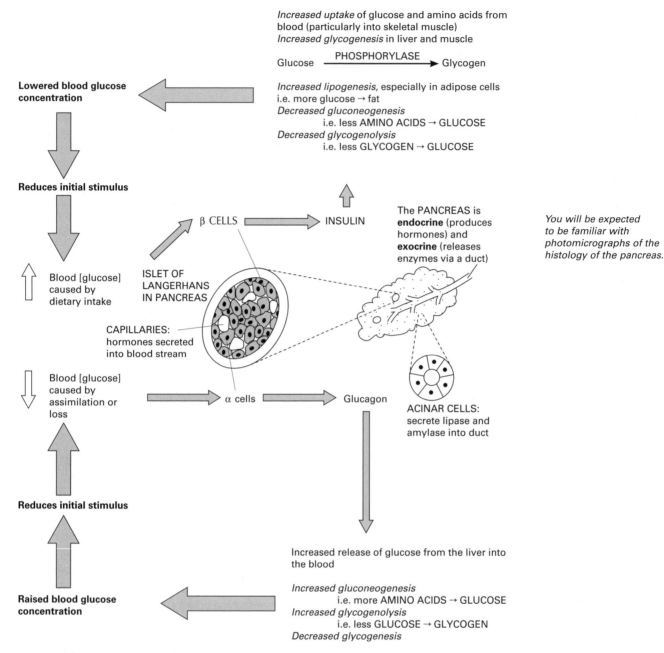

Increased uptake of glucose and amino acids from blood (particularly into skeletal muscle)
Increased glycogenesis in liver and muscle

$$Glucose \xrightarrow{PHOSPHORYLASE} Glycogen$$

Increased lipogenesis, especially in adipose cells
i.e. more glucose → fat
Decreased gluconeogenesis
        i.e. less AMINO ACIDS → GLUCOSE
Decreased glycogenolysis
        i.e. less GLYCOGEN → GLUCOSE

Lowered blood glucose concentration

Reduces initial stimulus

β CELLS ⟹ INSULIN

The PANCREAS is **endocrine** (produces hormones) and **exocrine** (releases enzymes via a duct)

You will be expected to be familiar with photomicrographs of the histology of the pancreas.

ISLET OF LANGERHANS IN PANCREAS

Blood [glucose] caused by dietary intake

CAPILLARIES: hormones secreted into blood stream

Blood [glucose] caused by assimilation or loss

α cells ⟹ Glucagon

ACINAR CELLS: secrete lipase and amylase into duct

Reduces initial stimulus

Increased release of glucose from the liver into the blood

Increased gluconeogenesis
        i.e. more AMINO ACIDS → GLUCOSE
Increased glycogenolysis
        i.e. less GLUCOSE → GLYCOGEN
Decreased glycogenesis

Raised blood glucose concentration

Hormones of the pancreas regulate blood glucose concentration by negative feedback

How insulin secretion is controlled

Beta cells have a resting potential across their cell membranes – they are negative inside compared with their surroundings. This resting potential is generated by potassium ion (K^+) channels which pump out K^+ ions and are sensitive to ATP.

If the concentration of glucose in the plasma rises, glucose enters the beta cell and is respired. This results in an increased ATP concentration in the cell, which closes the K^+ channels. The negative resting potential is removed and the membrane is depolarized. Calcium (Ca^{2+}) channels in the membrane are then open; Ca^{2+} ions pass into the cell and they stimulate the release of insulin from the beta cell by exocytosis.

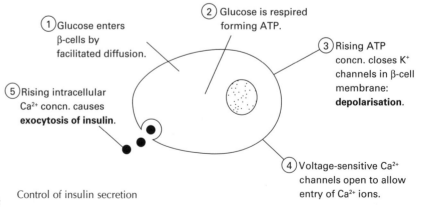

① Glucose enters β-cells by facilitated diffusion.

② Glucose is respired forming ATP.

③ Rising ATP concn. closes K^+ channels in β-cell membrane: **depolarisation**.

⑤ Rising intracellular Ca^{2+} concn. causes **exocytosis of insulin**.

④ Voltage-sensitive Ca^{2+} channels open to allow entry of Ca^{2+} ions.

Control of insulin secretion

Genetically engineered human insulin

Type 1 diabetes is treated by injections of insulin. In the past this insulin was extracted from the pancreases of other species (cows or pigs). There are slight differences in the structure of the insulin protein molecule between species and some patients' immune systems produce antibodies against the injected insulin, neutralising its actions and resulting in inflammatory responses at injection sites. There were also fears of long-term complications arising from the regular injection of a foreign substance. Today insulin is produced by recombinant DNA technology, i.e. inserting the human insulin gene into the *E. coli* bacterial cell, to produce an insulin that is chemically identical to naturally produced human insulin.

Human insulin also acts more quickly and for a shorter time than pig insulin, and people whose response to pig insulin is declining have a better result with human insulin.

Stem cells

Stem cells are defined by their
- potential to differentiate into different cell types
- ability to divide continually by mitosis, producing new generations of cells.
- **Embryonic stem cells** derive from totipotent and pleuripotent cells.
- **Adult stem cells** usually derive from multipotent cells.

	all	**totipotent cells**	development of the embryo
potential to differentiate into different cell types	most	↓ **pleuripotent cells**	
	limited	↓ **multipotent cells**	↓

Fact file

Adult stem cells have been manipulated *in vitro* to produce embryonic-like stem cells. The adult cells are dedifferentiated. Since embryos are not the source of these stem cells, the technique may help to make embryonic stem cell therapy more acceptable.

The science behind stem cell therapy

In adult humans some stem cells remain in the body. For example, adult stem cells in the marrow of the leg bones and arm bones give rise to different types of blood cell. If injury or disease damages a tissue, then its stem cells divide and differentiate into new cells, repairing the damage. Stimulating embryonic or adult stem cells to multiply and differentiate *in vitro* has the potential of making unlimited supplies of different types of cell available to be transplanted into people whose tissues are so damaged as to be beyond self-repair. This is called **stem cell therapy**.

The potential of stem cell therapy

Remember that embryonic stem cells can differentiate into many types of cell other than adult stem cells. This makes them ideal for different stem cell therapies. However their use is controversial because the current methods of obtaining embryonic stem cells destroys the embryos from which the cells are sourced.

Using adult stem cells to repair damaged tissue is less controversial because embryos are not destroyed. Also, sourcing the cells from the person who is to receive treatment removes the risk of rejection when the cells are transplanted back into the person.

Potential use of stem cells to treat diabetes

Instead of injecting insulin, doctors have attempted to cure diabetes by injecting patients with pancreatic islet cells grown in culture from human embryonic stem cells. Steroid immunosuppressant therapy is needed to prevent rejection of the cells, and this damages beta cells which may eventually no longer produce insulin.

Research into this approach shows promise. Remaining challenges are to ensure that transplanted cells are not rejected by the recipient, and that they differentiate and become integrated so that they function effectively within the host's pancreas.

The **heart rate** is measured as the number of ventricular contractions per minute. Usually we think of the contractions as the heart beat.

Recall from your work at *AS level* that the sino-atrial node (SAN) is located in the wall of the right atrium of the heart. Tissue in the SAN, called the **pacemaker**, determines a spontaneous average heart rate of about 100 beats per minute. The average heart rate of a healthy person at rest is about 72 beats per minute. Nervous and hormonal control of the heart rate accounts for the difference in the figures. The diagram represents connections between the brain, different receptors, and the SAN. Together, they help to control the heart rate.

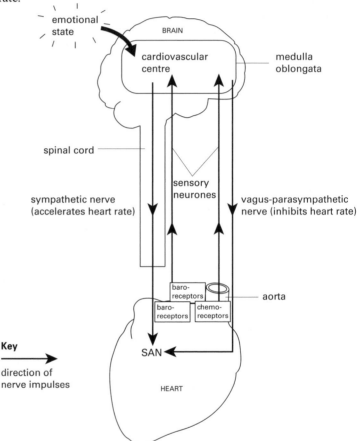

Notice:

- The **cardiovascular centre** is located in the part of the brain called the **medulla oblongata**.
- The **sympathetic nerve** and **vagus nerve** lead from the cardiovascular centre to the pacemaker in the SAN. The nerves are part of the **autonomic nervous system** which controls the involuntary (unconscious) actions of the body. *Recall* that the autonomic nervous system consists of sympathetic nerves and parasympathetic nerves.
- The sympathetic nerve passes from the cardiovascular centre to the pacemaker via the spinal cord.
- The vagus nerve is a parasympathetic nerve. It passes from the cardiovascular centre directly to the pacemaker.
- **Baroreceptors** are located in a swelling of the carotid artery called the **carotid sinus**, and in the walls of the heart.
- **Chemoreceptors** are located in the
  - glandular tissue of the **carotid body** near to the carotid artery, branching from the aorta
  - aorta
  - walls of the heart
  - medulla oblongata of the brain

Nervous control of heart rate

Baroreceptors are a type of mechanoreceptor. They detect changes in pressure. Here they detect changes in blood pressure. The chemoreceptors are sensitive to changes in blood pH and concentrations of carbon dioxide and oxygen.

This sensory information passes as nerve impulses via sensory neurones to the cardiovascular centre in the medulla oblongata of the brain. The information is processed in the medulla oblongata, stimulating the sympathetic nerve and the vagus nerve. *Recall* that these nerves pass from the medulla oblongata to the pacemaker in the SAN.

Nerve impulses transmitted to the pacemaker by the

- sympathetic nerve *accelerate* heart rate
- vagus nerve *inhibit* heart rate

The effect of the sympathetic nerve on the pacemaker opposes the effect of the vagus nerve. We say that the effects are **antagonistic**. A person's actual heart rate depends on the balance of activity between the nerves.

Recall that the spontaneous average heart rate is about 100 beats per minute. This means that the average resting heart rate of 72 beats per minute is set by the activity of the vagus nerve.

Heart rate varies

Age, gender, and fitness affect the resting heart rate. On average at rest, a child's heart rate is faster than an adult's, an adult woman's heart rate is faster than that of a man's, and the heart rate of an endurance athlete is likely to be slowest of all.

Exercise affects heart rate irrespective of age, gender, or fitness and increases in proportion to the intensity of the activity and the individual's oxygen uptake. It leads to

- an increase in concentration of carbon dioxide in the blood
- a decrease in blood pH
- The changes are detected by the chemoreceptors.
- The sensory information passes to the cardiovascular centre.
  - As a result the sympathetic nerve is stimulated and nerve impulses pass to the pacemaker.
  - As a result the heart rate increases.
  - As a result the output of blood from the heart (cardiac output) increases.
  - As a result more blood with its load of carbon dioxide passes to the lungs.
  - As a result more carbon dioxide is exhaled.
  - As a result the concentration of carbon dioxide in the blood decreases.
  - As a result stimulation of the pacemaker by nerve impulses from the sympathetic nerve decreases.
  - As a result heart rate returns to normal.

Notice that the process is an example of **negative feedback** which regulates the concentration of carbon dioxide in the blood (and its pH, and indirectly its oxygen concentration) and therefore the heart rate.

Hormonal control of heart rate

Hormones also affect heart rate. **Adrenaline** is released by the sympathetic nervous system in response to a person's emotional state (excitement, fear).

- Adrenaline increases the heart rate.
  - As a result more oxygen and nutrients reach the muscles.
  - As a result more glucose is available, enabling muscle cells to respire more rapidly and produce more ATP.
  - As a result more energy resources are available, enabling muscles to contract more vigorously.
  - As a result the individual is able to respond quickly to the stimulus which triggered the sequence of events in the first place.

Because of the consequences of its activity, adrenaline is sometimes called the 'flight or fight' hormone.

Questions

1. How are the sympathetic nerve and vagus nerve antagonistic to one another?

2. Briefly describe the role of sensory receptors in the control of the heart rate.

3. What is the result of the effect of adrenaline on the heart rate?

Excretion is the process by which waste products of metabolism and other non-useful substances are eliminated from an organism.

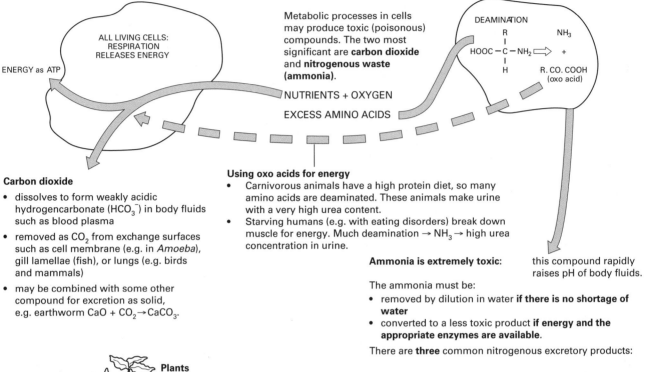

Metabolic processes in cells may produce toxic (poisonous) compounds. The two most significant are **carbon dioxide** and **nitrogenous waste (ammonia)**.

ALL LIVING CELLS: RESPIRATION RELEASES ENERGY

ENERGY as ATP

NUTRIENTS + OXYGEN

EXCESS AMINO ACIDS

DEAMINATION

$$HOOC - \overset{\overset{\displaystyle R}{|}}{\underset{\underset{\displaystyle H}{|}}{C}} - NH_2 \implies \quad \overset{NH_3}{\underset{R.\,CO.\,COOH}{+}}$$
(oxo acid)

Carbon dioxide

- dissolves to form weakly acidic hydrogencarbonate (HCO_3^-) in body fluids such as blood plasma
- removed as CO_2 from exchange surfaces such as cell membrane (e.g. in *Amoeba*), gill lamellae (fish), or lungs (e.g. birds and mammals)
- may be combined with some other compound for excretion as solid, e.g. earthworm $CaO + CO_2 \rightarrow CaCO_3$.

Using oxo acids for energy

- Carnivorous animals have a high protein diet, so many amino acids are deaminated. These animals make urine with a very high urea content.
- Starving humans (e.g. with eating disorders) break down muscle for energy. Much deamination $\rightarrow NH_3 \rightarrow$ high urea concentration in urine.

Ammonia is extremely toxic: this compound rapidly raises pH of body fluids.

The ammonia must be:

- removed by dilution in water **if there is no shortage of water**
- converted to a less toxic product **if energy and the appropriate enzymes are available**.

There are **three** common nitrogenous excretory products:

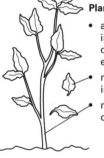

Plants

- autotrophic, so balance intake of nutrients to demand – little need for excretion
- may excrete some wastes in falling leaves
- may store some wastes in old ('heartwood') xylem.

Product	Toxicity	Water required for excretion	Energy required for production
Ammonia	High	High	Low
Urea	Medium	Medium	High
Uric acid	Low	Low	High

Testing urine samples for pregnancy and for misuse of anabolic steroids

Products of metabolism, or substances taken in with the diet, may be filtered by the kidney and excreted in the urine.

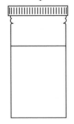

Pregnancy: Human chorionic gonadotrophin (HCG) is released by the developing placenta. It is detected by antibodies on a plastic testing strip dipped in to urine: a dye indicates the result. HCG in the urine of a pregnant woman binds to blue-tagged HCG antibodies in the dipstick. These bound antibodies rise up the dipstick. In the test window is a row of immobilised antibodies which bind to the blue-tagged HCG complex. A blue line shows a positive response (HCG present, confirming pregnancy). The control window has a row of immobilised antibodies which bind to blue-tagged antibodies in the absence of HCG. A blue line here confirms the test was carried out correctly.

YES!

Steroid abuse: Testosterone or its products can be detected in urine.

Glucose: Blood sugar at high levels in diabetic people means that sugar can be detected in the urine, and is detected by a Clinistix (Glucose oxidase strip).

1.13 The liver

Structure of the liver

The liver is the largest gland, weighing between 1.0 and 2.3 kg. It is situated in the upper abdomen, just beneath the diaphragm, and has four lobes. On the posterior surface is the **portal fissure** where various structures – hepatic portal vein, hepatic artery, hepatic vein, bile duct, lymph vessels – together with sympathetic and parasympathetic nerve fibres, enter or leave the gland.

You will be expected to be familiar with photographs of the gross structure and histology of the liver.

Blood supply to the liver

Hepatic vein – blood is deoxygenated and returns to general circulation.

Hepatic portal vein carries blood from the digestive tract, including branches from stomach, pancreas, spleen, ileum, colon and rectum.

Hepatic artery – blood is fully oxygenated and at normal arterial pressure. Liver has high oxygen demand for aerobic respiration.

Intralobular (central) vein: unites with others to form the **hepatic vein** which returns 'modified' blood to the general circulation.

Each lobe is composed of many tiny **lobules** which are the functional units of the liver.

Liver cord or **string of hepatocytes** in contact with blood flowing through sinusoid.

Liver sinusoid is a blood vessel with incomplete walls which allows 'input' blood to come into close contact with hepatocytes and Küppfer cells.

Interlobular vein is a branch of the portal vein and delivers blood, with varying solute concentration, from the gut.

Interlobular artery is a branch of the hepatic artery delivering oxygenated blood to the liver.

Each lobule has pairs of columns of **hepatocytes** radiating from a central vein. Between two pairs of columns are **sinusoids** and each sinusoid is serviced by a branch of the hepatic portal vein, a branch of the hepatic artery and a tributary of the bile duct. Each group of cells with its associated blood supply is called a **liver acinus**.

Bile canaliculus: unites with others to form the bile duct to take away the liver secretion, bile, from between the strings of hepatocytes.

Hepatocytes: liver cells

The **hepatocyte** or liver cell has microvilli to increase surface area for absorption from the percolating blood in the sinusoids, prominent Golgi body for the preparation of cellular products for secretion, rough endoplasmic reticulum for the synthesis and intracellular transport of numerous proteins, glycogen granules or fat globules to act as energy reserves, and numerous mitochondria to generate ATP for the liver's numerous metabolic reactions. The plasma membrane has numerous transporter proteins as well as insulin receptors.

Although all hepatocytes perform all liver functions there is some localisation of function. For example, cells closest to the portal capillaries are most active in gluconeogenesis and glycogenesis and in oxidative ATP formation whilst those closest to the central vein are most active in fat synthesis, glycolysis and drug metabolism. It is these perivenous cells which are most prone to damage following paracetamol overdose.

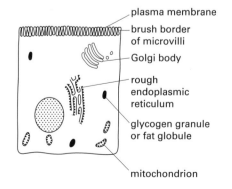

plasma membrane

brush border of microvilli

Golgi body

rough endoplasmic reticulum

glycogen granule or fat globule

mitochondrion

Liver structure and function: a single sinusoid

Branch of the hepatic portal vein (interlobular vein)
Delivers blood from the absorptive regions of the gut carrying a **variable concentration** of the soluble products of digestion.

Branch of hepatic artery (interlobular artery)
Delivers oxygenated blood at high pressure – hepatocytes have a high oxygen demand and are very vulnerable to hypoxia. Also delivers lactate from anaerobic respiration in skeletal muscle.

Küppfer cell
A fixed phagocytic cell which removes old ('effete') red blood cells. Iron is stored as ferritin, globin → amino acid pool and pyrrole rings are excreted as bile pigments.

Storage of minerals and vitamins

principally **iron** (as ferritin), some potassium, copper and trace elements.

mainly the **fat-soluble** vitamins A, D, E. **Small** amounts of B_{12}, C. Synthesis of vitamin A from carotene.

Bile ductile
Takes away **bile** to be stored in the gall bladder. Bile is 90% water + **bile salts** (aid emulsification of fats) + **bile pigments** (an excretory product) + **cholesterol + salts**. Release is triggered by CCK-PZ from wall of duodenum.

Polar bears store so much vitamin A in their liver that this organ becomes toxic to other animals! Huskies beware!

Blood filled channel

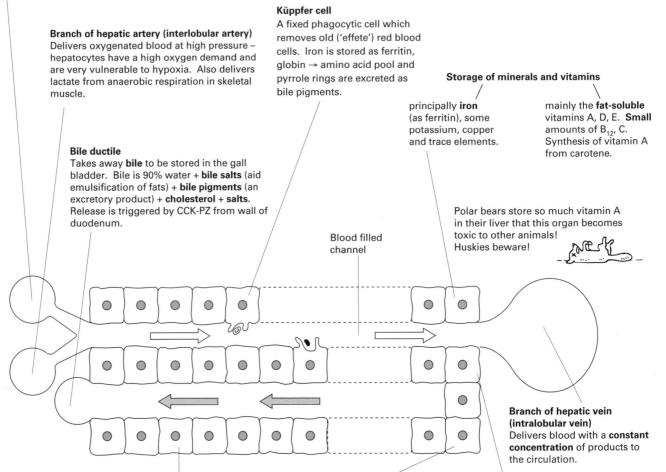

Branch of hepatic vein (intralobular vein)
Delivers blood with a **constant concentration** of products to the circulation.

Detoxification of compounds taken in with food (e.g. alcohol), produced during metabolism or released by pathogens:
- often involves **oxidation** e.g. alcohol → ethanal
- breakdown products may cause liver damage (cirrhosis by ethanal, membrane damage by paracetamol)
- includes catalase breakdown of peroxide from respiration.

Hormone removal
Rapid inactivation of testosterone and aldosterone. More gradual breakdown of insulin, glucagon, thyroxine, cortisone, oestrogen and progesterone.

Liver cells: correct the composition of the blood plasma as it flows past.

Heat production
High metabolic rate and considerable energy consumption make the liver the main heat-producing organ in the body. Metabolic rate and hence heat production is under the control of thyroxine.

Formation of urea: the ornithine cycle

Urea is formed from excess amino acids inside hepatocytes.

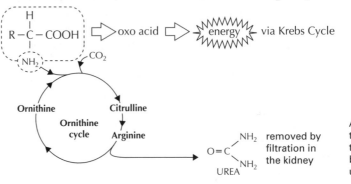

Ammonia is too toxic to be safely transported in the bloodstream, but urea is not.

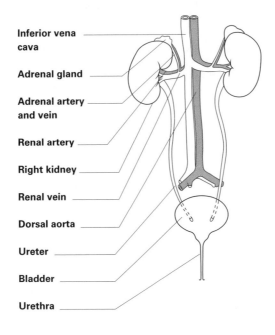

Inferior vena cava

Adrenal gland

Adrenal artery and vein

Renal artery

Right kidney

Renal vein

Dorsal aorta

Ureter

Bladder

Urethra

Renal circulation: handles approx. 1200 cm³ of blood per minute ~ about 25% of cardiac output.

Interlobular artery: delivers blood at high pressure to the glomerular capillaries.

Interlobular vein: drains blood away from the glomerular filtration units and from the Loops of Henle.

Medulla: has a striated appearance due to the presence of the Loop of Henle, the collecting duct and the vasa rectae.

Renal papilla: apex of the pyramid from which the ends of the collecting ducts deliver the urine.

Ureter: propels urine from the pelvis to the bladder.

Kidney cortex: has a smooth texture and is the site of the Bowman's Capsules and the glomeruli, together with the proximal and distal convoluted tubules and their associated blood supply.

Renal pyramid: one of 5–12 triangular structures which make up the medullary region.

Pelvis: urine collects here from tips of renal papillae.

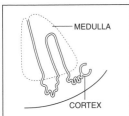

MEDULLA

CORTEX

Single nephron: (much enlarged) to show position in medulla and cortex.

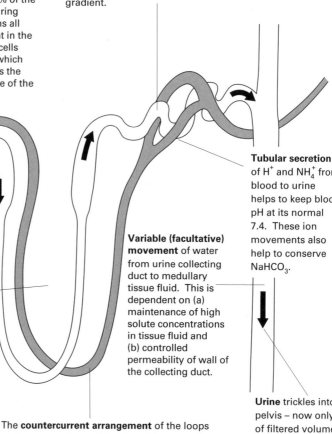

Ultrafiltration (glomerular filtration) occurs in the renal capsule: the selective structure is the **basement membrane of the glomerulus.** Water and solutes of relative molecular mass less than 68 000 form the filtrate.

Glomerular filtrate is formed at above 125 cm³ per minute in humans (this represents about 20% of the plasma delivered during that time). It contains all the materials present in the blood except blood cells and most proteins, which are too large to cross the basement membrane of the glomerulus.

Active controlled reabsorption of Na⁺ occurs in the distal convoluted tubule. This is followed by the osmotic movement of an equivalent volume of water down the water potential gradient.

Selective reabsorption occurs in the proximal tubule. Solutes are selectively moved from the filtrate to the plasma by **active transport** and water follows by osmosis. Almost all glucose and amino acids, and high but variable amounts of ions, are reabsorbed here. Since water follows by osmosis (about 80% of the filtrate volume), and is not controlled by the proximal tubule, this is referred to as **obligatory water reabsorption.**

The different permeability properties of the two limbs of the Loop of Henle, together with their counterflow arrangement, allows a **countercurrent multiplication** to generate a high solute concentration in the tissue fluid of the medulla. The highest solute concentrations are generated deep in the medulla.

Variable (facultative) movement of water from urine collecting duct to medullary tissue fluid. This is dependent on (a) maintenance of high solute concentrations in tissue fluid and (b) controlled permeability of wall of the collecting duct.

Tubular secretion of H⁺ and NH₄⁺ from blood to urine helps to keep blood pH at its normal 7.4. These ion movements also help to conserve NaHCO₃.

The **countercurrent arrangement** of the loops of the vasa rectae and the sluggish movement of blood through them means that few ions are removed and the high solute concentration generated by the Loop of Henle is maintained.

Urine trickles into kidney pelvis – now only 1% of filtered volume, high concentrations of urea, creatinine and variable ion concentration. Typically about 1.5 dm³ per day.

The functions of the nephron

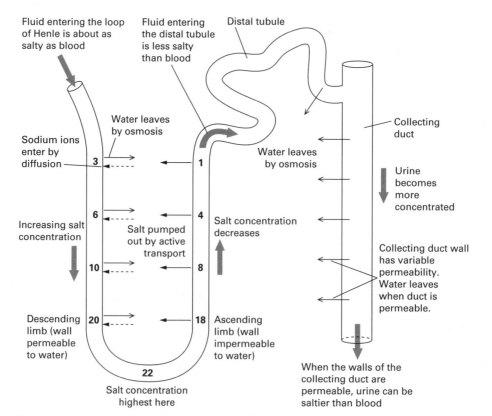

Countercurrent multiplier mechanism in the loop of Henle. The figures show relative salt concentrations

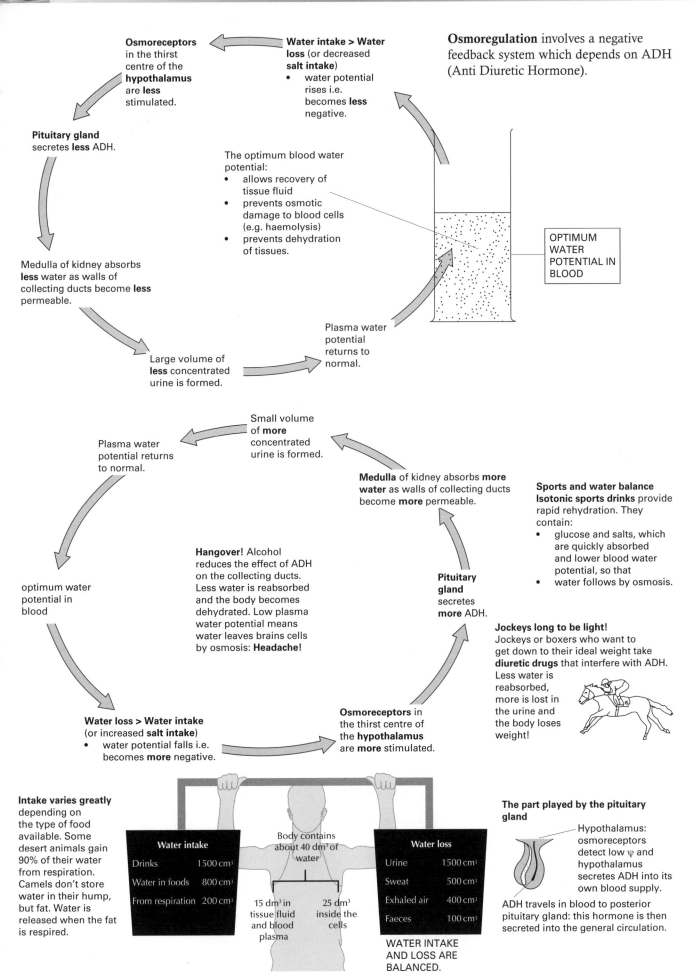

Osmoregulation involves a negative feedback system which depends on ADH (Anti Diuretic Hormone).

Osmoreceptors in the thirst centre of the **hypothalamus** are **less** stimulated.

Water intake > Water loss (or decreased **salt intake**)
• water potential rises i.e. becomes **less** negative.

Pituitary gland secretes **less** ADH.

The optimum blood water potential:
• allows recovery of tissue fluid
• prevents osmotic damage to blood cells (e.g. haemolysis)
• prevents dehydration of tissues.

OPTIMUM WATER POTENTIAL IN BLOOD

Medulla of kidney absorbs **less** water as walls of collecting ducts become **less** permeable.

Large volume of **less** concentrated urine is formed.

Plasma water potential returns to normal.

Small volume of **more** concentrated urine is formed.

Plasma water potential returns to normal.

Medulla of kidney absorbs **more water** as walls of collecting ducts become **more** permeable.

Sports and water balance
Isotonic sports drinks provide rapid rehydration. They contain:
• glucose and salts, which are quickly absorbed and lower blood water potential, so that
• water follows by osmosis.

optimum water potential in blood

Hangover! Alcohol reduces the effect of ADH on the collecting ducts. Less water is reabsorbed and the body becomes dehydrated. Low plasma water potential means water leaves brains cells by osmosis: **Headache!**

Pituitary gland secretes **more** ADH.

Jockeys long to be light!
Jockeys or boxers who want to get down to their ideal weight take **diuretic drugs** that interfere with ADH. Less water is reabsorbed, more is lost in the urine and the body loses weight!

Water loss > Water intake (or increased **salt intake**)
• water potential falls i.e. becomes **more** negative.

Osmoreceptors in the thirst centre of the **hypothalamus** are **more** stimulated.

Intake varies greatly depending on the type of food available. Some desert animals gain 90% of their water from respiration. Camels don't store water in their hump, but fat. Water is released when the fat is respired.

Body contains about 40 dm³ of water

Water intake	
Drinks	1500 cm³
Water in foods	800 cm³
From respiration	200 cm³

15 dm³ in tissue fluid and blood plasma

25 dm³ inside the cells

Water loss	
Urine	1500 cm³
Sweat	500 cm³
Exhaled air	400 cm³
Faeces	100 cm³

WATER INTAKE AND LOSS ARE BALANCED.

The part played by the pituitary gland

Hypothalamus: osmoreceptors detect low ψ and hypothalamus secretes ADH into its own blood supply.

ADH travels in blood to posterior pituitary gland: this hormone is then secreted into the general circulation.

If one or both kidneys fail then dialysis is used or a transplant performed to keep urea and solute concentration in the blood constant.

Kidney failure may result from injury or from long-term disease of the kidneys. The kidneys can no longer filter toxins and waste products from the blood. There may be protein or blood in the urine.

Symptoms of kidney failure include high levels of urea in the blood, which can lead to:

- vomiting and diarrhoea, causing dehydration
- weight loss
- more or less frequent urination than normal, and increased volume of urine
- blood in the urine.

There is also a build-up of phosphates in the blood which may cause

- itching
- bone damage
- broken bones fail to rejoin
- muscle cramps (caused by low levels of calcium).

Increased potassium in the blood may cause:

- abnormal heart rhythms
- muscle paralysis.

Failure of kidneys to remove excess fluid may cause:

- swelling of the legs, ankles, feet, face and/or hands
- shortness of breath due to extra fluid on the lungs.

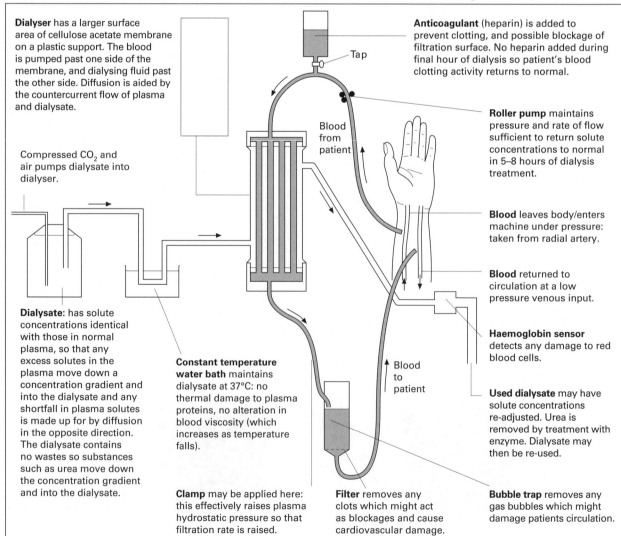

Dialyser has a larger surface area of cellulose acetate membrane on a plastic support. The blood is pumped past one side of the membrane, and dialysing fluid past the other side. Diffusion is aided by the countercurrent flow of plasma and dialysate.

Compressed CO_2 and air pumps dialysate into dialyser.

Dialysate: has solute concentrations identical with those in normal plasma, so that any excess solutes in the plasma move down a concentration gradient and into the dialysate and any shortfall in plasma solutes is made up for by diffusion in the opposite direction. The dialysate contains no wastes so substances such as urea move down the concentration gradient and into the dialysate.

Constant temperature water bath maintains dialysate at 37°C: no thermal damage to plasma proteins, no alteration in blood viscosity (which increases as temperature falls).

Clamp may be applied here: this effectively raises plasma hydrostatic pressure so that filtration rate is raised.

Tap

Blood from patient

Blood to patient

Filter removes any clots which might act as blockages and cause cardiovascular damage.

Anticoagulant (heparin) is added to prevent clotting, and possible blockage of filtration surface. No heparin added during final hour of dialysis so patient's blood clotting activity returns to normal.

Roller pump maintains pressure and rate of flow sufficient to return solute concentrations to normal in 5–8 hours of dialysis treatment.

Blood leaves body/enters machine under pressure: taken from radial artery.

Blood returned to circulation at a low pressure venous input.

Haemoglobin sensor detects any damage to red blood cells.

Used dialysate may have solute concentrations re-adjusted. Urea is removed by treatment with enzyme. Dialysate may then be re-used.

Bubble trap removes any gas bubbles which might damage patients circulation.

Kidney transplantation may be necessary as renal dialysis is inconvenient for the patient and costly.

Kidney transplants have a high success rate because:

1 the vascular connections are simple

2 live donors may be used, so very close blood group matching is possible

3 because of 2 there are fewer immuno-suppression-related problems in which the body's immune system reacts against the new kidney.

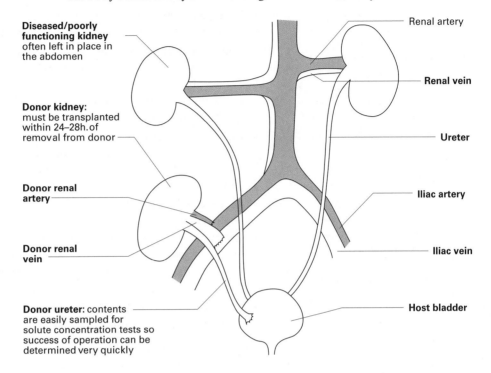

Diseased/poorly functioning kidney often left in place in the abdomen

Donor kidney: must be transplanted within 24–28h. of removal from donor

Donor renal artery

Donor renal vein

Donor ureter: contents are easily sampled for solute concentration tests so success of operation can be determined very quickly

Renal artery

Renal vein

Ureter

Iliac artery

Iliac vein

Host bladder

Fact file

- **Autotrophs** such as green plants can synthesize their own food from inorganic substances, using light (the process of photosynthesis) or chemical energy (chemosynthesis).
- **Heterotrophs** such as animals cannot manufacture their own food. They obtain food and energy by taking in organic substances, usually plant or animal matter.
- In **respiration** organic substances are broken down by a series of reactions that usually involve oxygen (aerobic respiration), to release energy and produce carbon dioxide.
- The substrates for respiration in both autotrophs and heterotrophs are the organic substances built up by autotrophic nutrition. So both plants and animals depend on photosynthesis for respiration.

Plants fix carbon dioxide and water, forming complex organic molecules. Light is the source of energy which drives the reactions. The reactions are the components of **photosynthesis**. It is a reduction process in which carbon dioxide is reduced by hydrogen derived from water. The sequence of events is shown in the diagram.

In **light harvesting**, light energy is captured by chlorophyll and other light absorbing pigments. The two stages of photosynthesis take place in the chloroplast:

- **Light-dependent reactions** – captured light energy
  - splits water into hydrogen ions (H^+) and oxygen: the process is called **photolysis**
  - is converted into the bond energy of ATP: the process is called **photophosphorylation**
- **Light-independent reactions** – hydrogen ions (H^+) released by photolysis combine with carbon dioxide, reducing it. Triose sugar is formed.

Recall the adaptations of a leaf which maximize the rate of photosynthesis. The diagram reminds you of the structure of a leaf and its tissues.

- Palisade cells beneath the upper surface of the leaf form the major photosynthetic tissue of most plants.
- Chloroplasts stream in the cytoplasm of palisade cells to the region where light is brightest. The process is called **cyclosis**.
- The elongated shape of palisade cells enables them to act as light tubes, transmitting light to tissues deep within the leaf.
- Spaces in the leaf tissue enable carbon dioxide and water vapour (the raw materials of photosynthesis) to circulate freely within the leaf.

Adaptations of the chloroplast

The **outer envelope** of the chloroplast separates it from the cytoplasm of the plant cell, isolating the reactions of photosynthesis from the rest of the cell's metabolism.

The reactions of the light-dependent reactions take place on the thylakoid membranes of the chloroplast, while the light-independent stage is located in the stroma.

- The disc-shaped **thylakoid membranes** contain chlorophyll. They are stacked in piles to form the **grana**; this structure increases the surface area of the thylakoid membranes so allowing photosynthesis to happen quickly in sunlight.
- The **stroma** contains the enzymes of the light-independent stage.
- **Starch grains** in the stroma store the carbohydrate produced in photosynthesis.

You will be expected to be familiar with electron micrographs of the chloroplast.

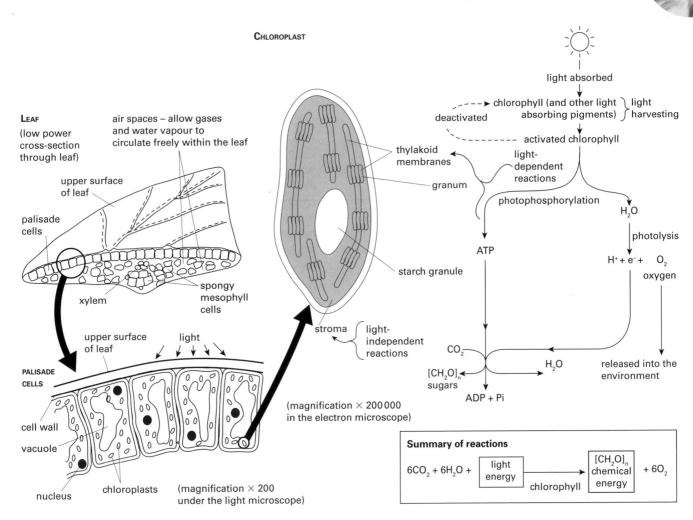

Leaf cells and chloroplasts. Chlorophyll and other light-absorbing pigments cover the thylakoid membranes of each granum. Their large surface maximizes the capture of light. The light-dependent reactions occur on the thylakoid membranes; the light-independent reactions occur in the stroma.

Questions

1 Summarize the different stages of photosynthesis.

2 Explain how the structure of a chloroplast maximizes the rate of photosynthesis.

1.18 Light harvesting

Chlorophyll and other photosynthetic pigments

A **pigment** is a substance that absorbs light. Chlorophyll is a pigment. It absorbs light of all colours (wavelengths) except green, which is mostly reflected. It is not a single compound but a mixture of several.

- **Chlorophyll** a – the **primary** photosynthetic pigment, so called because it is directly responsible for the conversion of light energy to the bond energy of ATP. The diagram shows its structure.

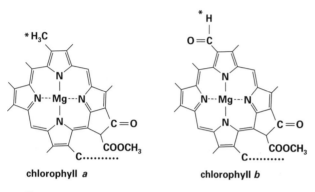

chlorophyll a   **chlorophyll b**

Key

C········· = long hydrocarbon tail

- ○ *Notice* that the 'head' of the molecule consists of a magnesium ion held in a ring system called a **porphyrin ring**. The structure is similar to that of haem, except that a magnesium ion substitutes for an iron ion.
- ○ *Notice* that a long hydrocarbon 'tail' is attached to the ring. The 'tail' is part of the thylakoid membrane of the chloroplast; the 'head' lies on its surface.
- ○ *Notice* the alternate double and single bonds of the porphyrin ring. When illuminated, the bonds switch back and forth as singles and doubles conducting electrons. We say that chlorophyll is **photoexcitable**.

- **Chlorophyll** b – differs from chlorophyll a in that an aldehyde group (–CHO) replaces the methyl group (–CH$_3$) in the position marked * on the porphyrin ring illustrated. It is an **accessory** pigment, so called because it transfers energy (photoexcited electrons) to chlorophyll a.
- **Carotenoids** – a group of compounds which are also **accessory** pigments. There are two categories: *carotenes* which are derived from vitamin A, and xanthophylls.

Carotenoids range in colour from pale yellow, orange, to deep red and are responsible for the colours of many fruits and flowers.

The accessory pigments absorb light at wavelengths different from those absorbed by chlorophyll a.

- As a result the energy of most of the spectrum of visible light blue → red is made available for photosynthesis.

Each photosynthetic pigment absorbs particular wavelengths of light. The band of wavelengths is the **absorption spectrum** of the pigment. The **action spectrum** is a measure of how effective different wavelengths of light are in bringing about biological processes – in this case photosynthesis.

The graphs illustrate the absorption spectra for different photosynthetic pigments and the action spectrum for photosynthesis. *Notice* the close correlation between the absorption spectrum for each of the photosynthetic pigments and the action spectrum for the rate of photosynthesis by the plant as a whole. The correspondence between the two spectra is evidence that the photosynthetic performance of the plant is related to the wavelengths of light absorbed by the pigments.

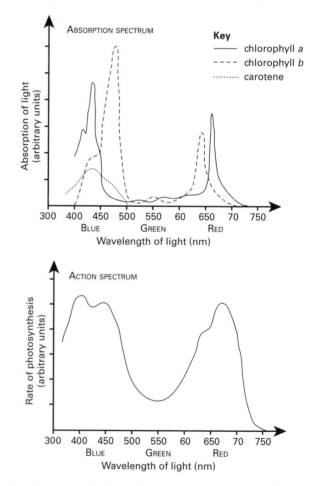

Absorption spectra for different plant pigments and the action spectrum for photosynthesis

Light harvesting

Recall that primary chlorophyll and accessory photosynthetic pigments are part of the thylakoid membranes. They are packed into units called **photosystems**. Each unit contains between 250 and 400 molecules of pigment and traps photons of light energy.

Recall also the difference between primary and accessory photosynthetic pigments. Once absorbed by a molecule of one of the accessory pigments, a photon passes between the other accessory pigment molecules. Eventually it transfers to and is absorbed by a molecule of a special form of chlorophyll *a* – the primary pigment. The molecule is called the **reaction centre chlorophyll molecule**. It is the **reaction centre** of the photosystem. The diagram shows you the idea.

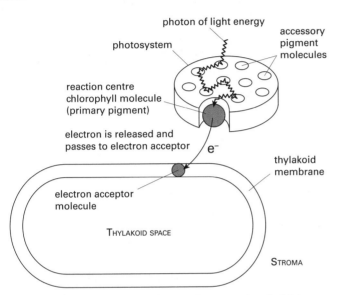

An antenna complex absorbing photons. The process is called **light harvesting**. It involves numerous pigment molecules for each photon absorbed.

When a reaction centre chlorophyll molecule absorbs a photon, the photon's energy excites an electron.

- As a result the energy level of the electron is raised.
- As a result the electron is released and transfers to a primary electron acceptor (electron acceptor 1).
- As a result the chlorophyll molecule is oxidized and positively charged.

What happens next? Reading on will help you to answer the question.

Fact file

Like the electron transport chain of the inner mitochondrial membrane, the electron transport chain linking the primary electron acceptor 1 to photosystem 1 (PSI) consists of different proteins which transfer electrons along the chain in a series of redox reactions.

- Energy released by the redox reactions enables protons to be removed from the chloroplast stroma across the thylakoid membrane into the space enclosed by the membrane.

- As a result protons accumulate in the thylakoid space.

- As a result a proton gradient develops across the thylakoid membrane.

- Protons accumulated in the thykaloid space flow down the proton gradient (chemiosmosis) from the space through channel proteins and ATP synthase to the stroma.

- As a result energy is released.

- As a result ADP combines with Pi. The reaction is catalysed by ATP synthase. ATP is formed.

Questions

1. Why is chlorophyll called a pigment?
2. What is the difference between the absorption spectrum and action spectrum of a pigment?
3. Briefly describe the functional relationship between the different photosynthetic pigments of a photosystem.

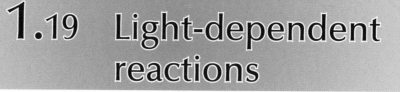

There are two different photosystems, each containing a slightly different version of the reaction centre chlorophyll molecule from the other:

- **photosystem I (PSI)** in which the peak absorption spectrum of the reaction centre chlorophyll molecule is 700 nm (P_{700})
- **photosystem II (PSII)** in which the peak absorption spectrum of the reaction centre chlorophyll molecule is 680 nm (P_{680})

The two systems require light and work together at the same time. The diagram and its checklist are your guide to the details of the light-dependent reactions, which take place in the thylakoid membranes.

Stage ❶

Photons of light energy are harvested by PSII and trapped by P_{680} forming the reaction centre of the photosystem. Its electrons are raised to a higher energy level, released and captured by electron acceptor 1. Now oxidized, the loss of electrons by the P_{680} of the reaction centre is repaid by electrons gained from the photolysis of water. *Remember* that photolysis also produces protons (H^+) and that oxygen gas is released.

Stage ❷

Electrons transfer from electron acceptor 1 along an electron transport chain to PSI. The transfer of electrons is by a series of redox reactions similar to that of the electron transport chain of the inner mitochondrial membrane. Energy is released enabling ADP and inorganic phosphate (Pi) to combine, forming ATP. Light energy has therefore been converted to and stored as chemical bond energy in molecules of ATP. The process is called **non-cyclic photophosphorylation** as its reactions follow a linear metabolic pathway.

Stage ❸

Photons of light energy are harvested by PSI and trapped by P_{700} forming the reaction centre of photosystem. Its electrons are raised to a higher energy level. The electrons are released and captured by electron acceptor 2. Now oxidized, the loss of electrons by the P_{700} of the reaction centre is repaid by electrons gained from the chain of electron carriers described in Stage 2.

Stage ❹

Some electrons from electron acceptor 2 pass back to PSI by the chain of electron carriers described in Stage 2. Another molecule of ATP is generated. Because electrons are recycled the process is called **cyclic photophosphorylation**.

Stage ❺

Electrons transfer from electron acceptor 2 along a chain of electron carriers to **nicotinamide adenine dinucleotide phosphate (NADP)**. They combine with the protons (H^+) released by photolysis (see Stage 1) and NADP is reduced to NADPH.

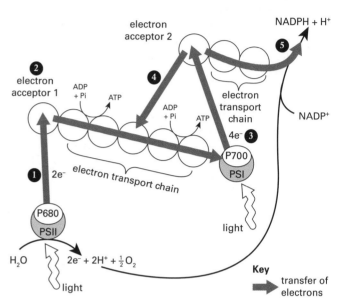

Light-dependent reactions of photosynthesis

The role of water in the light-dependent stage

In the light-dependent stage water is broken down into hydrogen and oxygen using light energy. This oxygen is released as a by-product at the end of photosynthesis. The hydrogen goes on to enter the light-independent stage.

Questions

1 What is the difference between non-cyclic photophosphorylation and cyclic phosphorylation?

2 What are the sources of electrons transferred to electron acceptor 1 and electron acceptor 2?

3 How is photosystem I different from photosystem II?

1.20 Light-independent reactions

The **light-independent** reactions of photosynthesis take place in the stroma of chloroplasts. They occur whether or not light is available. The sequence of reactions is sometimes called the **Calvin cycle**. In this cycle, the end product of the reactions regenerates the starting substance of the process. The diagram and its checklist are your guide to the details of the light-independent reactions.

Stage ❶

Carbon dioxide in solution diffuses through the plasma membrane, through the cytoplasm and through the membrane surrounding the chloroplasts of photosynthetic cells. In the stroma it combines with the 5-carbon compound **ribulose bisphosphate (RuBP)**, producing an unstable 6-carbon molecule which immediately splits into two 3-carbon compound molecules of **glycerate 3-phosphate (GP)**. The reaction is catalysed by the enzyme **ribulose bisphosphate carboxylase (rubisco)** which is located on the surface of the thylakoid membranes.

Stage ❷

GP is reduced to **triose phosphate (TP)**, by NADPH produced in the light-dependent reactions. The reaction is endothermic and driven by the ATP produced during photophosphorylation. $NADP^+$ is regenerated and is available to accept more protons (H^+) released by photolysis during the light-dependent reactions.

Stage ❸

The combination of two molecules of TP produces a molecule of glucose (a 6-carbon sugar). The process is a reversal of glycolysis. The combination of many glucose molecules forms starch.

Stage ❹

The majority of TP molecules are recycled, combining in a variety of reactions, regenerating RuBP. Overall 5 molecules of TP → 3 molecules of RuBP. The process uses ATP produced by photophosphorylation during the light-dependent reactions.

Products of photosynthesis

All plant cells use sugar, including TP and glucose, as the starting point for the synthesis of other carbohydrates and lipids. The addition of nitrogen forms amino acids and therefore proteins, as well as other nitrogen containing compounds. These different syntheses are endothermic and are driven by the ATP produced by cellular respiration.

The role of carbon dioxide in the light-independent stage

In the Calvin cycle carbon dioxide is fixed (converted to carbohydrate in the plant). The carbon dioxide combines with hydrogen, which originates from water, to form the carbohydrate glucose.

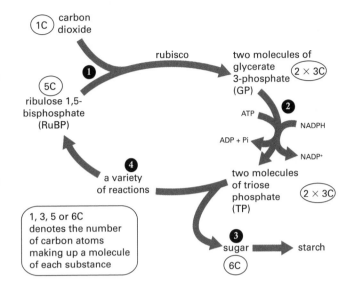

The light-independent reactions of photosynthesis

Qs and As

Q What are the biochemical outcomes of the light-dependent reactions?

A • *Light energy is converted into chemical energy.*

• *The photolysis of water releases protons (H^+), electrons (e^-), and oxygen gas.*

• *ATP and NADPH + H^+ are produced and available in the light-independent reactions where carbon dioxide is reduced, forming triose sugar.*

Questions

1 How is nicotinamide adenine dinucleotide phosphate (NADP) regenerated in the light-independent reactions?

2 What is meant by the statement that 'the combination of two molecules of TP produces a molecule of glucose by reverse glycolysis'?

1.21 Limits on the rate of photosynthesis

The rate of photosynthesis determines the mass of triose sugar produced in a given time. It is affected by supplies of **carbon dioxide** and **water**, **temperature** and the **intensity of light**. These factors are called **limiting factors** because if any one of them falls to a low level the rate of photosynthesis slows or stops – even if the other factors are in abundant supply. The greater the rate of photosynthesis the greater is the growth rate of plants. Therefore, if limiting factors slow the rate of photosynthesis then the growth rate of the plants affected also slows.

Light intensity and concentration of carbon dioxide

At constant temperature the rate of photosynthesis varies with the intensity of light. The top graph illustrates the relationship. For example:

- As dawn lightens the night sky, photosynthesis begins. *Notice* that the increase in the rate of photosynthesis is proportional to the increase in light intensity through the early morning. Light intensity is the limiting factor.
- As the intensity of light continues to increase, so too does the rate of photosynthesis. However the relationship is no longer directly proportional. Light is still a limiting factor but some other factor is limiting as well.
- Yet further increase in light intensity does *not* lead to further increase in the rate of photosynthesis. Light **saturation** is reached and the intensity of light is no longer a limiting factor. Some other factor is limiting the process.

Since temperature is constant, the 'other' limiting factor is carbon dioxide. The second graph makes the point.

- A low concentration of carbon dioxide limits the rate of photosynthesis whatever the light intensity. As the concentration of carbon dioxide increases, there is a proportional increase in the rate of photosynthesis.
- As the concentration of carbon dioxide continues to increase, so too does the rate of photosynthesis. However the relationship is no longer directly proportional. The concentration of carbon dioxide is still a limiting factor, but some other factor is limiting as well.
- Yet further increase in the concentration of carbon dioxide does not lead to further increase in the rate of photosynthesis and the concentration of carbon dioxide is no longer a limiting factor. Notice that at higher concentrations of carbon dioxide the rate of photosynthesis increases with increasing light intensity and then levels off. Now light intensity is the limiting factor.

Temperature

When light intensity is not limiting (high intensity), the rate of photosynthesis increases proportionately as temperature increases... over a limited range. However when light intensity is limiting (low intensity), the rate of photosynthesis does not increase with an increase in temperature. Why? *Remember* that the process of photosynthesis consists of two stages: light-dependent and light-independent. The light-dependent reactions are **photochemical**. Most photochemical reactions are not temperature sensitive. However the enzyme catalysed reactions of the light-independent reactions are. So, increase in temperature increases the rate of photosynthesis to the point when temperature is no longer a limiting factor. Conditions are optimal for enzyme activity.

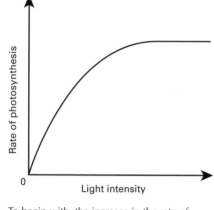

To begin with, the increase in the rate of photosynthesis is proportional to the increase in light intensity.

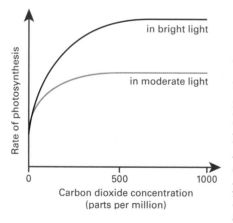

Carbon dioxide is a limiting factor

Compensation point

At night photosynthesis stops, so plants do not consume carbon dioxide but produce it because their cells continue to respire. At dawn photosynthesis begins, and some of the carbon dioxide produced through respiration is consumed. Less carbon dioxide therefore is released into the environment.

As light intensity increases, so too does the rate of photosynthesis, and less and less carbon dioxide is released. A point is reached where the carbon dioxide used by photosynthesis is balanced by the carbon dioxide produced by aerobic respiration. The *status quo* is called the **compensation point**.

Plants that grow in unshaded sunny habitats (**sun plants**) have a high compensation point. Buttercups are an example. For plants that flourish under the canopy of trees and other shaded habitats (**shade plants**) the compensation point is lower. Dogs' mercury and ivy are examples.

Notice in the graphs that the rate of photosynthesis of shade plants is maximal at a lower light intensity than for sun plants. Shade plants therefore would outgrow sun plants in shaded habitats. However, the maximum rate of photosynthesis for sun plants is greater than shade plants. Sun plants therefore would outgrow shade plants in sunny habitats.

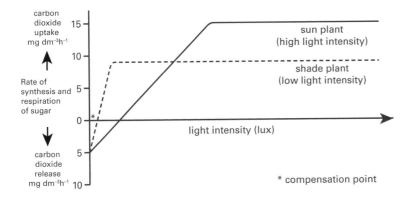

Greenhouses eliminate limiting factors

A greenhouse provides everything plants need for photosynthesis. Conditions are controlled so that the rate of photosynthesis is at a maximum.

The environment inside the greenhouse can be precisely controlled. Sensors linked to computers can monitor light intensity, temperature, moisture, and the concentration of carbon dioxide in the air. The computer processes the information and alters the controls to provide optimal growing conditions for the plants.

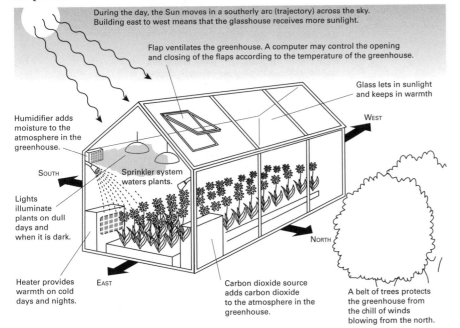

Plants grow quickly under glass because the greenhouse environment maximizes their rate of photosynthesis

Fact file

Why are limiting factors limiting?

- When carbon dioxide is in short supply, the rate of conversion of RuBP to GP in the light-independent reactions decreases.

- As a result RuBP builds up and TP is not produced.

- At low temperature, the number of collisions between enzyme/substrate molecules decreases.

- As a result the rate of the enzyme-catalysed reactions producing TP during the light-independent reactions decreases.

- At low light intensity, the production of ATP and NADPH in the light-dependent reactions decreases (stops).

- As a result GP is not converted into TP during the light-independent reactions.

- As a result GP builds up and RuBP is used up.

These effects of limiting factors reduce the rate of photosynthesis and therefore slow the rate of plant growth.

Questions

1 What is the compensation point?

2 Define a limiting factor.

3 Briefly discuss why greenhouses help to eliminate limiting factors.

Investigating limiting factors

Audus' photosynthometer can be used to measure the rate of photosynthesis while varying the temperature, light intensity, or carbon dioxide concentration.

Principle: bubbles of evolved gas are collected over a known period of time.

$$\text{Rate of photosynthesis} = \frac{\text{collected volume} / mm^3}{\text{time} / min}$$

$= mm^3 \, O_2$ evolved / min at a known temperature, $t°C$.

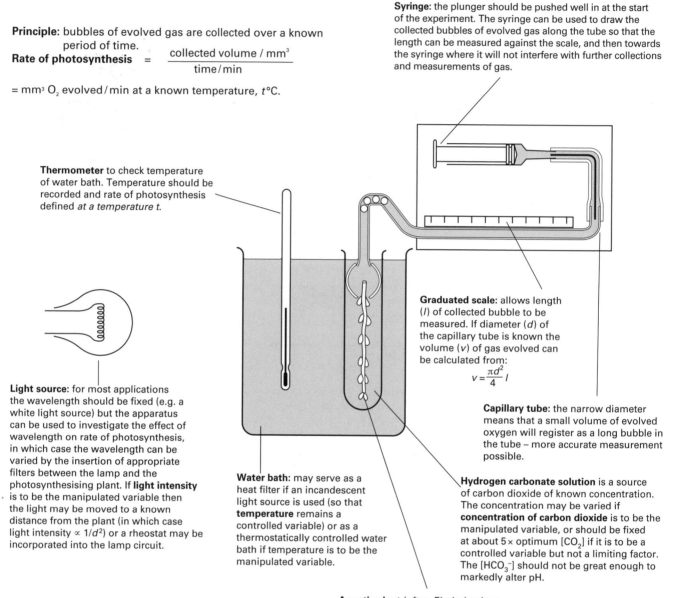

Syringe: the plunger should be pushed well in at the start of the experiment. The syringe can be used to draw the collected bubbles of evolved gas along the tube so that the length can be measured against the scale, and then towards the syringe where it will not interfere with further collections and measurements of gas.

Thermometer to check temperature of water bath. Temperature should be recorded and rate of photosynthesis defined *at a temperature t.*

Graduated scale: allows length (*l*) of collected bubble to be measured. If diameter (*d*) of the capillary tube is known the volume (*v*) of gas evolved can be calculated from:

$$v = \frac{\pi d^2}{4} l$$

Light source: for most applications the wavelength should be fixed (e.g. a white light source) but the apparatus can be used to investigate the effect of wavelength on rate of photosynthesis, in which case the wavelength can be varied by the insertion of appropriate filters between the lamp and the photosynthesising plant. If **light intensity** is to be the manipulated variable then the light may be moved to a known distance from the plant (in which case light intensity ∝ $1/d^2$) or a rheostat may be incorporated into the lamp circuit.

Capillary tube: the narrow diameter means that a small volume of evolved oxygen will register as a long bubble in the tube – more accurate measurement possible.

Water bath: may serve as a heat filter if an incandescent light source is used (so that **temperature** remains a controlled variable) or as a thermostatically controlled water bath if temperature is to be the manipulated variable.

Hydrogen carbonate solution is a source of carbon dioxide of known concentration. The concentration may be varied if **concentration of carbon dioxide** is to be the manipulated variable, or should be fixed at about 5× optimum [CO$_2$] if it is to be a controlled variable but not a limiting factor. The [HCO$_3^-$] should not be great enough to markedly alter pH.

Aquatic plant (often *Elodea*): when photosynthesising, releases oxygen from a cut end of the stem. The plant specimen can be induced to photosynthesise actively by prior illumination, and gentle aeration of the solution for about an hour before the experiment. Any adjustment of the manipulated variable (for example, a change in light intensity), should be followed by a short period (ca. 5 min) to allow the plant to re-adjust (it allows the system to equilibrate).

Problems:
1. Are **controlled variables** at their optimum?
2. Are all plant samples comparable – same size, age, activity?
3. Is apparatus reliable – clean, leakproof?
4. Are collected bubbles all oxygen, all produced by photosynthesis?

1.22 ATP

ATP is short for **adenosine triphosphate**. Its molecule consists of the sugar **ribose** to which is attached the base **adenine** and a chain of three **phosphate** groups. The combination of ribose and adenine forms the **adenosine** part of the molecule.

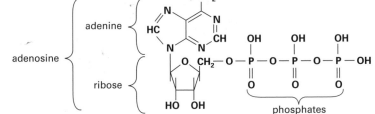

ATP is formed when a phosphate group (Pi) binds to a molecule of **ADP** (adenosine diphosphate).

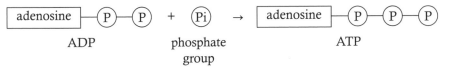

The reaction is:

- **a phosphorylation** – a type of reaction where a phosphate group is added to another molecule
- **endothermic** – a type of reaction which absorbs energy
- catalysed by the enzyme ATP **synthase**

The energy driving the synthesis of ATP comes from:

- *light* in the light-dependent reactions of photosynthesis
- *sugars* (also lipids and proteins) which are oxidized in the reactions of cellular respiration

In each case the energy released is coupled with the phosphorylation of ADP, producing ATP.

Releasing energy from ATP

ATP is an immediate source of energy for biological processes. It is very soluble in water and often described as the *universal energy currency* found in the cells of all living organisms... from bacteria to oak trees to humans.

When a molecule of ATP combines with a water molecule, the bond binding the endmost phosphate group to the rest of the ATP molecule is broken. The reaction is a **hydrolysis**. During the reaction energy is consumed breaking the bond, but more energy is released as other bonds are formed. Overall the reaction is therefore **exothermic**.

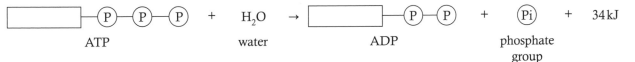

The hydrolysis of ATP is also catalysed by ATP synthase. This time the enzyme works in reverse to its activity catalysing the production of ATP. The energy released is coupled to and drives other biological processes including:

- **anabolic reactions** which result in the synthesis of **polymers** from building block units, for example monosaccharides → polysaccharides
- **active transport** of substances against their concentration gradients across cell membranes, for example
  - o the transfer of glucose from blood to liver cells
  - o the exchange of sodium ions (Na^+) and potassium ions (K^+) across the membrane of the axon of a nerve cell, generating an action potential
- **muscle contraction** during which muscle fibres shorten

A proportion of the energy from the hydrolysis of ATP is released as heat energy. In birds and mammals some of the heat helps to maintain a constant body temperature. The rest is transferred to the environment.

Respiration and ATP

Why do plants, animals, and micro-organisms all need to respire? In respiration organic substances are broken down by a series of reactions that usually involve oxygen (aerobic respiration), to release energy and produce carbon dioxide. This energy is in the form of ATP, a molecule that stores energy in cells. This energy can be released again from ATP to drive processes of life including metabolism, active transport, and (in animals) muscle contraction.

Questions

1 Name two sources of energy that drive the synthesis of ATP.

2 Why is adenosine triphosphate so-named?

3 Explain why the synthesis of ATP is described as an endothermic reaction.

Fact file

The term **metabolism** refers to all of the chemical reactions taking place in a cell.

- **Catabolic** reactions break down large molecules into smaller ones. The reactions of cellular respiration are an example.

- **Anabolic** reactions make (synthesize) large molecules from smaller ones. The synthesis of polysaccharides, lipids, and proteins are examples.

A metabolic pathway is a sequence (particular order) of reactions where a particular molecule is converted into another different one by way of a series of intermediate compounds.

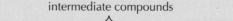

intermediate compounds

molecule A $\longrightarrow$ molecule B $\longrightarrow$ molecule C $\longrightarrow$ molecule D $\longrightarrow$ molecule E

Only small amounts of energy are released or taken in during the reaction. Cells are therefore not damaged. Reactions which require or release large amounts of energy would destroy cells.

Remember that a substance which an enzyme enables to react is called a **substrate**. The substance produced as a result of the reaction is the **product**.

If the product of one reaction of a metabolic pathway inhibits one of the enzymes preceding its formation, the product will act as an inhibitor of the whole pathway. The process is called **end-point inhibition**.

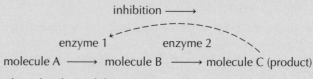

inhibition $\longrightarrow$

enzyme 1   enzyme 2

molecule A $\longrightarrow$ molecule B $\longrightarrow$ molecule C (product)

For example molecule C inhibits enzyme 1. The metabolic pathway as a whole therefore slows down or stops. End-point inhibition is an example of negative feedback – the mechanism of homeostasis. It controls the rates of reaction of a metabolic pathway.

Key words

Oxidation – a reaction in which a substance gains oxygen atoms or loses hydrogen atoms or electrons.

Reduction – a reaction in which a substance loses oxygen atoms or gains hydrogen atoms or electrons.

Electron – a particle with a negative charge that orbits the nucleus of an atom.

Coenzyme – a molecule which binds to an atom (or group of atoms) of one molecule and transfers the atom (or group of atoms) to another molecule. Coenzyme A transfers an acetyl group in the link reaction, while NAD and FAD are coenzymes that transfer hydrogen in a variety of redox reactions.

When glucose is oxidized completely in the presence of oxygen, the products are carbon dioxide and water. Energy is released. Overall the equation is

$$C_6H_{12}O_6(aq) + 6O_2(g) \rightarrow 6CO_2(g) + 6H_2O(l) \quad \Delta H_R -2880 \text{ kJ}$$

...where ΔH_R represents the energy change of the reaction at standard temperature and pressure.

If the energy change of the reaction were released all at once in cells, then the increase in temperature would destroy them. Instead, the energy is released step-by-step in a series of reactions in four stages: **glycolysis**, the **link reaction**, the **Krebs cycle**, and the **electron transport chain**.

Aerobic respiration refers to the link reaction, the Krebs cycle and the reactions of the electron transport chain because the oxygen which enables the reactions to take place comes from the air. The reactions of aerobic respiration occur in the mitochondria. Some of the energy released during the reactions is stored in molecules of ATP.

How much ATP is produced during aerobic respiration?

During aerobic respiration ATP is produced from two sources:

- **substrate-level phosphorylation** – a type of chemical reaction where ATP is produced by the direct transfer of a phosphate group to ADP from another reactive substance. It occurs in the presence or absence of oxygen.

- **oxidative phosphorylation** – electrons are released during the reactions of glycolysis, the link reaction, and the Krebs cycle. The electrons are accepted by the coenzymes **nicotinamide adenine dinucleotide (NAD)** and **flavine adenine dinucleotide (FAD)** which are reduced to NADH and $FADH_2$ respectively. Ultimately the electrons are transferred from reduced NAD and FAD to oxygen which serves as the final electron acceptor. During the transfer of electrons energy is released and ATP produced. Oxidative phosphorylation occurs *only* in the presence of oxygen.

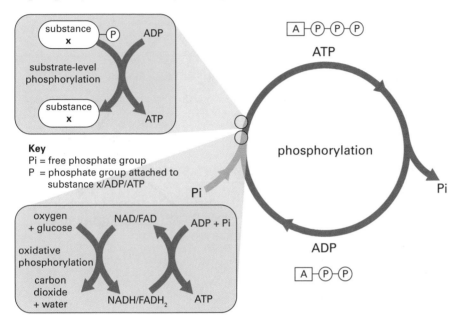

Key
Pi = free phosphate group
P = phosphate group attached to
     substance x/ADP/ATP

The ATP cycle

Fact file

Respiration in anaerobic conditions by

- yeast cells produces ethanol and carbon dioxide

- muscle cells produces lactic acid

The processes are examples of **fermentations**.

The table shows the number of molecules of ATP used and produced when a molecule of glucose is completely oxidized in the reactions of aerobic respiration.

Stage of aerobic respiration	ATP used	ATP produced by substrate-level phosphorylation	Number of reduced molecules of NAD and FAD formed	ATP produced by oxidative phosphorylation	Total production of ATP molecules
glycolysis	2	4	2NADH + H⁺	6	10
link reaction	0	0	2NADH + H⁺	6	6
Krebs cycle	0	2	6NADH + H⁺	18	24
			2FADH₂	4	
					40

Note that the net yield of ATP molecules per molecule of glucose is 38 because two molecules of ATP are used in glycolysis.

The net yield of 38 molecules of ATP following aerobic respiration represents about 40% of the potential energy contained in a molecule of glucose. The remaining 60% of energy is released as heat. In birds and mammals, the high rate of cellular respiration is the source of heat which enables them to maintain constant body temperature.

When oxygen is in short supply in cells we say that conditions are **anaerobic**. The electron transfer chain cannot work and the reactions of the Krebs cycle stop. Only the reactions of glycolysis take place. The table shows that 4 molecules of ATP are produced by substrate-level phosphorylation during glycolysis. However, 2 molecules are used so the net number of ATP molecules available following respiration in anaerobic conditions is only 2.

Fact file

In practice, the net yield of 38 ATP molecules produced during aerobic respiration is never reached. Losses occur because more ATP molecules are used than just those during glycolysis.

Questions

1 Summarize the different stages of aerobic respiration.

2 Why are fewer molecules of ATP produced during anaerobic respiration than during aerobic respiration?

3 What is the difference between substrate-level phosphorylation and oxidative phosphorylation?

Glycolysis literally means 'splitting sugars'. A molecule of glucose – a six carbon (hexose) sugar is split into two molecules of a three carbon (triose) sugar. Each molecule of triose sugar is converted into a molecule of pyruvic acid as a pyruvate ion.

$$CH_3COCOOH \rightarrow CH_3COCOO^- + H^+$$

pyruvic acid   pyruvate ion   hydrogen ion

Glycolysis is the first stage of cellular respiration. It occurs in the cytoplasm of cells whether or not oxygen is abundant (aerobic conditions) or in short supply (anaerobic conditions).

- **Aerobic conditions** – pyruvate ions pass to Krebs cycle *via* the link reaction.
- **Anaerobic conditions** – pyruvate ions undergo a process of fermentation. Two of the most common types are lactic acid fermentation (muscle cells) and alcohol fermentation (yeast cells).

Fact file

Glycolysis is sometimes called the **Embden-Meyerhof pathway** after the two German biochemists who, in the 1930s, worked out its details.

Checklist: glycolysis

- A glucose (hexose) molecule is phosphorylated to hexose phosphate. The reaction is endothermic and makes the glucose molecule more reactive. The energy comes from the hydrolysis of ATP providing a phosphate group which binds to the glucose molecule.
$$ATP \rightarrow ADP + Pi$$
Phosphorylation commits the glucose molecule to enter the sequence of reactions of glycolysis.

- Further phosphorylation occurs, forming hexose bisphosphate. The reaction is endothermic and the addition of another phosphate group makes the hexose more reactive. Another molecule of ATP is used up.

- The molecule of hexose bisphosphate splits into two molecules of phosphorylated triose sugar, which in turn are converted in a series of reactions to pyruvate ions (you are not expected to remember the details). The reactions are exothermic.
  - Four phosphate groups are transferred from the molecules of triose sugar to ADP, forming four molecules of ATP (two molecules for each of triose sugar) by **substrate-level phosphorylation**.
  - At the same time the triose sugars are oxidized. Two pairs of hydrogen atoms (one pair for each molecule of triose sugar) are released. The loss of hydrogen by each molecule of triose sugar is an example of a **dehydrogenation** reaction. Each pair of hydrogen atoms is transferred to a molecule of nicotinamide adenine dinucleotide (NAD), forming reduced NAD (NADH + H$^+$).

- The formation of pyruvate ions marks the completion of glycolysis.

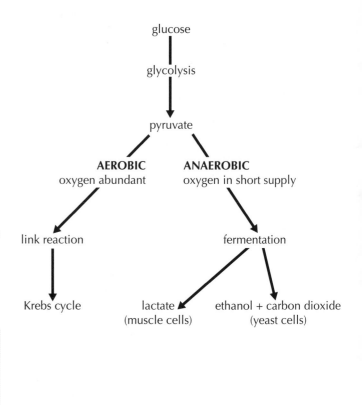

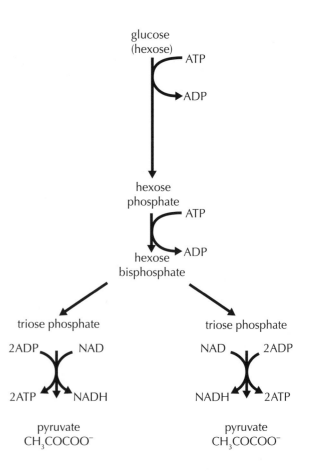

Checklist: fermentation

Remember that fermentation reactions take place in anaerobic conditions (when oxygen is in short supply). In the cytoplasm, pyruvate is converted into waste products which then may be removed from the cell.

In mammals

- Pyruvate produced by glycolysis in anaerobic conditions accumulates more rapidly than can be processed via the Krebs cycle.
- Oxidation of reduced NAD (NADH) transfers hydrogen atoms to pyruvate, reducing it.
    - As a result lactate ($CH_3CHOHCOO^- + H^+$) is formed.
    - As a result pyruvate does not accumulate during glycolysis.
    - As a result NAD is regenerated and is available to accept more hydrogen atoms.
- Regeneration of NAD is required if glycolysis is to continue. Without it glycolysis would stop.
- When more oxygen becomes available lactate either
    - undergoes reverse glycolysis synthesizing molecules of glucose, or
    - enters the Krebs cycle where it is oxidized, forming carbon dioxide and water.
- The amount of oxygen required to metabolize excess lactate is called the **oxygen debt**.

In yeast cells

- Pyruvate produced by glycolysis in anaerobic conditions accumulates more rapidly than can be processed via the Krebs cycle.
- The carboxyl group ($-COO^-$) of pyruvate is removed as carbon dioxide. The reaction is an example of **decarboxylation** catalysed by the enzyme **decarboxylase**.
    - As a result ethanal (CH_3CHO) is formed.
- Oxidation of reduced NAD (NADH) transfers hydrogen atoms to ethanal, reducing it.
    - As a result ethanol (CH_3CH_2OH) is formed.
    - As a result pyruvate does not accumulate during glycolysis.
    - As a result NAD is regenerated and available to accept more hydrogen atoms.
- Regeneration of NAD is required if glycolysis is to continue. Without it glycolysis would stop.

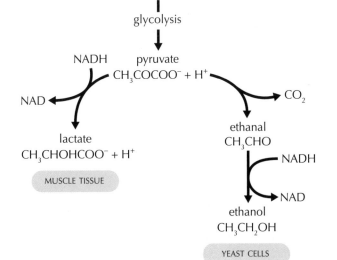

Fact file

Oxygen is rarely completely absent in cells or the environment. However its concentration may be too low to oxidize all of the pyruvate produced by glycolysis. The conditions are anaerobic and different fermentation reactions reduce the excess pyruvate, forming a variety of end products. Ethanol and lactate are common examples. Oxygen may be absent deep within rocks. We say the conditions are **hypoxic**. Some types of bacteria are found in rocks and can only live in hypoxic conditions. Oxygen poisons them.

Questions

1 At the start of glycolysis a glucose molecule is phosphorylated.
   a Explain the meaning of the statement.
   b What is the outcome of the reaction?
2 How is nicotinamide adenine dinucleotide (NAD) regenerated when
   a muscle cells
   b yeast cells
   respire anaerobically?
3 What is the oxygen debt?

Respiratory substrates

Here we are considering the respiration of glucose. However, if glucose is in short supply then protein and fat can be respired instead, as alternative **respiratory substrates**. Amino acids and fatty acids can both be converted to acetyl CoA and so enter the Krebs cycle.

The energy released by the respiration of a respiratory substrate depends on the amount of hydrogen in the molecule. Lipids with their long chains of fatty acids contain relatively larger proportions of hydrogen atoms than carbohydrates, so generate more energy per gram (about 39 kJ/g) than do carbohydrates. Proteins have a similar number of hydrogen atoms to carbohydrates so generate a similar amount of energy per gram (about 16 kJ/g).

Key words

Decarboxylation – a reaction where a carboxyl group ($-COO^- + H^+$) is removed from a substance. Carbon dioxide is produced.

Dehydrogenation – a reaction where hydrogen atoms are removed from a substance.

Redox reaction – involves the passage of electrons from one substance to another. The substance losing electrons is oxidized; the substance gaining electrons is reduced (hence the name 'redox'). Energy is transferred from one side of the reaction to the other.

Remember that for each molecule of glucose entering glycolysis, two molecules of pyruvic acid (as pyruvate ions) are produced. In aerobic conditions the pyruvate ions are actively transported into mitochondria. Their uptake consumes ATP.

In a mitochondrion the pyruvate ions enter the link reaction, which couples glycolysis with the reactions of the Krebs cycle. The diagram and its checklist are your guide to the details of the link reaction and the Krebs cycle.

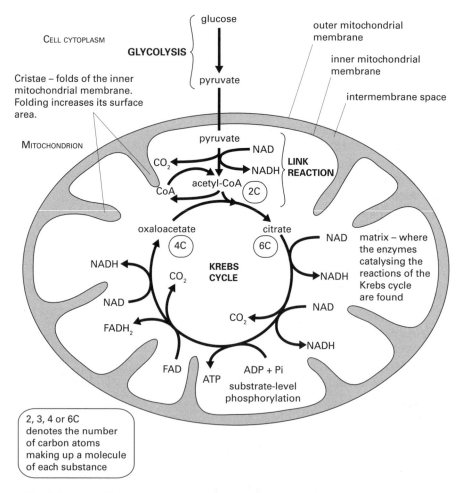

Checklist: the link reaction and the Krebs cycle

- Pyruvate passes into a mitochondrion where it is oxidized by the removal of hydrogen atoms to NAD, which is reduced to NADH. The removal of hydrogen atoms is another example of a dehydrogenation reaction catalysed by the enzyme **dehydrogenase**.

- A molecule of carbon dioxide is also released. The release is another example of a decarboxylation reaction catalysed by the enzyme **decarboxylase**.

As a result 3-carbon pyruvate is converted to a 2-carbon acetyl group.
- The acetyl group formed combines with coenzyme A (CoA), producing **acetyl coenzyme A** (abbreviated as **acetyl CoA**).

The formation of acetyl CoA couples glycolysis with the Krebs cycle: the so-called link reaction. This reaction takes place in the mitochondrial matrix.

- The acetyl group containing 2 carbon atoms carried by CoA is transferred to the 4-carbon compound **oxaloacetic acid** (as oxaloacetate ions).

As a result the 6-carbon compound **citric acid** (as citrate ions) is formed.

- A cyclic sequence of reactions follows during which 2 carbon atoms, each from a different compound in the sequence, are removed in the form of carbon dioxide (CO_2). Each removal is yet another example of a decarboxylation reaction catalysed by the enzyme decarboxylase.

Also pairs of hydrogen atoms are removed, each pair from a particular compound in the sequence of reactions. Each removal is yet another example of a dehydrogenation reaction catalysed by the enzyme dehydrogenase.

The hydrogen atoms either combine with NAD reducing it to NADH or FAD reducing it to $FADH_2$.

- ATP is produced by **substrate-level phosphorylation**.
- The formation of oxaloacetate marks the completion of the Krebs cycle, which like the link reaction occurs in the mitochondrial matrix.

The next sequence of reactions begins with the reaction of oxaloacetate with acetyl CoA, forming citrate.

What happens next?

Hydrogen atoms removed from compounds produced by the breakdown of food molecules (e.g. glucose) during glycolysis, the link reaction and the Krebs cycle are transferred ultimately as electrons (e^-) and protons (H^+) to oxygen, forming water. The transfer of electrons is by a series of **redox** reactions along a chain of electron acceptor molecules called the **electron transport chain**. The molecules of the electron transport chain are coenzymes and proteins which are associated with the inner mitochondrial membrane.

During the transfer of electrons energy is released. The energy enables protons to be pumped from the matrix of the mitochondrion, across the inner mitochondrial membrane, into the space between the inner mitochondrial membrane and the outer mitochondrial membrane. Protons accumulate in this intermembrane space. Their return flow to the matrix is facilitated by diffusion down their concentration gradient. Energy is released. The process is called **chemiosmosis**. The energy released enables ADP and inorganic phosphate (Pi) to combine, forming ATP.

Mitochondria: structure and function

The cigar shape of the mitochondrion gives a large surface area and short diffusion distance for the transport of oxygen, pyruvate, carbon dioxide, and ATP.

The outer mitochondrial membrane separates the organelle from the cytoplasm, forming a compartment where the reactions of respiration can occur. This membrane is permeable to pyruvate, oxygen, ATP, and many other molecules.

Inside are two further subdivided compartments: the matrix and the space between the inner and outer mitochondrial membranes.

In the matrix are the enzymes that catalyse the link reaction and the many reactions of the Krebs cycle.

The inner membrane is highly folded to form shelves called cristae, greatly increasing the surface area of the inner membrane for the reactions of the electron transport chain which generate ATP. It has structures called stalked particles which are the site of oxidative phosphorylation.

The intermembrane space has a high proton concentration generated by the electron transport chain of the inner membrane. This concentration is maintained by the fact that the inner membrane is not permeable to protons.

Fact file

The Krebs cycle is named after the biochemist Sir Hans Krebs who in the 1940s unravelled its sequence of reactions at Oxford University. The cycle is also known as the citric acid cycle or the **tricarboxylic acid (TCA) cycle**. Earlier in 1932, Krebs reported results of research which later helped other scientists to work out the reactions of the ornithine cycle whereby ammonia combines with carbon dioxide to form urea.

Questions

1 What is the difference between a dehydrogenation and decarboxylation reaction?

2 What is the role of acetyl CoA in aerobic respiration?

3 In which organelle of a cell and in what part of the organelle do the reactions of the Krebs cycle take place?

The synthesis of ATP by chemiosmosis is called **oxidative phosphorylation** because:

- the combination of ADP and Pi is a phosphorylation reaction
- oxygen is required as the final electron acceptor of the electron transport chain of proteins and coenzymes (labelled **A** to **D** and **x** and **y** respectively in the diagram below).

The diagram is your guide to the details of electron transfer and proton pumping which results in the oxidative phosphorylation of ADP forming ATP.

Transferring electrons

- During glycolysis, the link reaction, and the Krebs cycle, oxidation reactions remove hydrogen atoms from different compounds as electrons and protons.
- The reactions are coupled to the electron transport chain by the electron acceptors NAD and FAD.
- When NAD and FAD accept electrons they are reduced to NADH and FADH$_2$.
- NADH transfers electrons to protein **A** in the diagram of the electron transport chain.
  - As a result of the transfer NADH is oxidized to NAD and protein **A** is reduced.
- Electrons are transferred from reduced protein **A** to protein **C** by way of coenzyme **x**.
  - As a result reduced protein **A** is re-oxidized and protein **C** reduced.

- Electrons are also transferred from FADH$_2$ as part of protein **B** to protein **C** by way of coenzyme **x**.
  - As a result FADH$_2$ is oxidized to FAD and protein **C** is further reduced.
- Electrons are then transferred from reduced protein **C** to protein **D** by way of coenzyme **y**.
  - As a result reduced protein **C** is re-oxidized and protein **D** reduced.
- Electrons from reduced protein **D** are transferred with protons to oxygen. Molecules of water form and reduced protein **D** is re-oxidized, marking the end reaction of the electron transfer chain.

Pumping protons

- The redox reactions of the electron transport chain release energy which enables proteins **A**, **C**, and **D** to pump protons from the mitochondrial matrix across the inner mitochondrial membrane into the mitochondrial intermembrane space.
  - As a result protons accumulate in the intermembrane space. (*Remember* that protons carry a positive charge (H$^+$).)
  - As a result a proton gradient develops across the inner mitochondrial membrane.
- The potential difference across the inner mitochondrial membrane is −200 mV (the charge on the matrix side of the membrane is more negative than the side of the membrane facing the intermembrane space).
- The difference in charge results in an **electrochemical gradient** (of protons) and represents a store of energy.

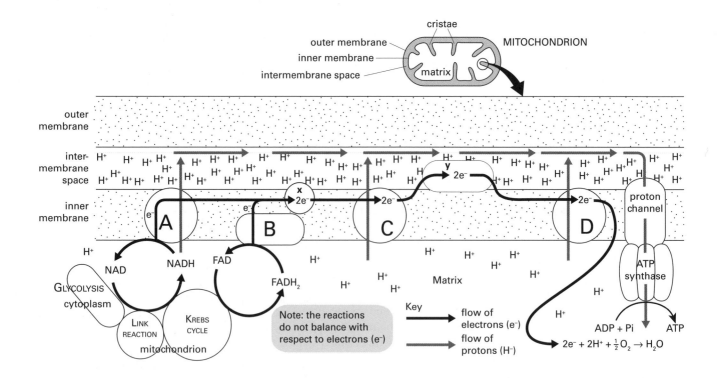

Fact file

The tendency of substances in solution to accept electrons is measured as their **redox potential**. *Remember* that when a substance accepts electrons it is reduced. The substance which donates the electrons is oxidized. Solutions with a high redox potential will tend to accept electrons from solutions with a lower redox potential.

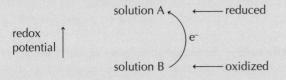

The sequence of the electron transport chains of the inner mitochondrial membrane and the thylakoid membranes of a chloroplast arises because of the difference in redox potentials between the carrier proteins. The higher its redox potential, the more likely is a carrier protein to accept electrons; the lower its redox potential the more likely is a carrier protein to donate electrons. For example **A → D** and **x** and **y** symbolizes the carrier proteins and coenzymes respectively associated with the inner mitochondrial membrane:

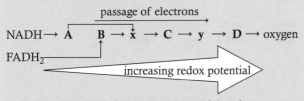

Oxygen is the final electron acceptor of the electron transport chain. Its redox potential is greater than protein **D**. The redox potential of protein **D** is greater than coenzyme **y**... and so on.

Synthesizing ATP

- Proton channels consisting of different proteins are part of the inner mitochondrial membrane.
- Each pore consists of
  - channel proteins which pass from the side of the membrane facing the intermembrane space to the matrix side of the membrane
  - the enzyme **ATPsynthase**, which is connected by a stalk to the channel proteins and projects into the matrix of the mitochondrion
- Protons accumulated in the intermembrane space flow down the **proton gradient** from the space through the channel proteins and ATPsynthase to the matrix (chemiosmosis).
  - As a result energy is released.
  - As a result ADP combines with Pi. The reaction is catalysed by ATPsynthase. ATP is formed (oxidative phosphorylation).

Evidence for chemiosmosis

- The proton gradient between the intermembrane space and the mitochondrial matrix can be measured, as it corresponds to a pH gradient.
- The ATPsynthase enzyme has been isolated, and will produce ATP in a proton gradient even if no electron transport is going on.
- Uncouplers are chemicals which uncouple electron transport and production of the proton gradient. These chemicals prevent the build-up of the proton gradient, and also prevent ATP synthesis.

Fact file

Recall the table on page 51 which sets out the number of ATP molecules generated when one molecule of glucose is oxidized to molecules of carbon dioxide and water during aerobic respiration. In theory, the oxidation of a molecule of NADH to NAD releases enough energy to charge the electrochemical gradient of protons (see **pumping protons**) with enough potential to generate 3 ATP molecules.

Similarly the oxidation of a molecule of $FADH_2$ to FAD releases energy with the potential of generating 2 ATP molecules. These are the values used in the table on page 51. However, it seems that NADH and $FADH_2$ generate only about 2.5 ATP and 1.5 ATP respectively, because not all of the energy stored in the proton gradient is available to generate ATP.

Qs and As

Q Cyanide is a poison. It binds with protein D of the electron transport chain shown in the diagram, preventing the transport of protons (H^+) from the mitochondrial matrix into the intermembrane space. Explain why a person who ingests (takes in) cyanide might die.

A *Your answer should include some (and preferably all) of the following points:*
- *a proton gradient does not develop*
- *protons do not flow through the channel proteins and ATPsynthase*
- *ATP is not formed by oxidative phosphorylation*
- *the Krebs cycle and the link reaction stop*
- *ATP is only formed by substrate-level phosphorylation during glycolysis*
- *lactate builds up*

Questions

1. What is the meaning of the term redox potential?
2. Summarize the process and the outcome of the process of chemiosmosis.

Genes and enzymes

Genes determine the composition not only of structural proteins such as keratin, but also of enzymes, most of which are globular proteins. Enzymes control the synthesis of carbohydrates, lipids, nucleic acids, and all other organic molecules in the cell. Enzymes are the key to how the sequence of bases in DNA in the nucleus controls all the inherited characteristics of organisms.

Recall that a **gene** is a section of **DNA** carrying information which enables a cell to combine amino acid units in the correct order forming a polymer (*or* part of a polymer) of peptide/polypeptide/protein. The information is carried in the sequence of bases on the nucleotides which make up the section of DNA forming a gene. The bases are adenine (A), thymine (T), cytosine (C), and guanine (G). The more amino acid units combined, the larger the polymer; peptide → polypeptide → protein. From now on *polypeptide* will include peptides/proteins.

Recall also that the **genetic code** is the sequence of bases of all of the genes of a cell and the information each gene carries. It works in the same way in the cells of all living things and is

- **a triplet code** – for each gene the information needed to assemble one amino acid unit in its correct place in a polypeptide molecule is contained in a sequence of three bases… the triplet is called a **codon**

- **non-overlapping** – the base of a triplet specifying the position of a particular amino acid unit does *not* contribute to specifying the positions of other amino acid units. So…

AGG	CCA	TAG	ACT	AAG	… and so on
codon	codon	codon	codon	codon	
↓	↓	↓	↓	↓	
amino acid 1	**amino acid 2**	**amino acid 3**	**amino acid 4**	**amino acid 5**	

Base A of this codon contributes to specifying the position of amino acid 1 in the polypeptide chain… …but does *not* contribute to specifying the position of the other amino acids 2–5 (likewise G, C, T)

- **degenerate** – the position of nearly all amino acids in a polypeptide molecule are specified by *more* than one codon.

Qs and As

Q How can a cell synthesize many more different polypeptides than there are genes which make up its genetic code?

A *Often, the synthesis of a polypeptide is the result of the combination of activity of two or more genes. The number of combinations is many more than the number of genes themselves.*

The nucleic acid family

Recall that a molecule of DNA is made up of two strands each consisting of many nucleotides joined together by condensation reactions. The strands coil round each other forming a **double helix** and are joined by hydrogen bonds between the bases **A T C** and **G**.

- A only bonds with T.
- G only bonds with C.

The arrangement is called **complementary base** pairing.

Recall also that the strands are **anti-parallel** to each other.

- One strand runs from carbon atom 3′ of one nucleotide to carbon atom 5′ of the next nucleotide in line… to carbon atom 3′ of the next nucleotide… and so on.

- Its partner strand runs 5′ → 3′ → 5′ … and so on in the opposite direction.

There are different types of RNA. The sequence of nucleotides (and therefore bases) of one of the strands of the section of DNA acts as a template (pattern) against which RNA is synthesized. For example …

Messenger RNA (mRNA) is synthesized in the cell nucleus. Its molecules are each usually single stranded, each strand consisting of many nucleotides joined together by condensation reactions. The sequence of nucleotides (and therefore bases) forming a strand of RNA is a *complement* of the sequence of nucleotides (and therefore bases) of the strand of the section of DNA which is the template against which the RNA strand forms.

Transfer RNAs (tRNA) tRNAs are smaller molecules than mRNA. The diagram represents the shape of a molecule of tRNA in the form of a cloverleaf. The shape is held by the hydrogen bonding of complementary bases within the molecule.

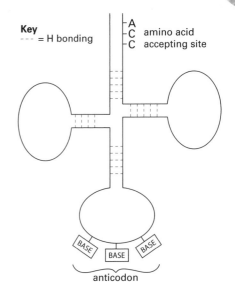

Key
--- = H bonding

A
C   amino acid
C   accepting site

anticodon

A transfer RNA (tRNA) molecule

• Notice in the diagram that an amino acid is attached at one end of the molecule which ends in a sequence of CCA. The sequence forms the **amino acid accepting site**. Transfer RNAs act as carriers of amino acids during polypeptide synthesis. Each type of amino acid is carried by its own type of tRNA.

• *Remember* that there are at least 20 different amino acids. There are, therefore, at least 20 different types of tRNA.

• Notice also the three bases which form the part of the molecule of tRNA labelled the anticodon. These bases are complementary to the mRNA codon encoding the amino acid it carries. There are as many different arrangements of the sequence of the anticodon as there are amino acids.

Ribosomal RNA (rRNA) combines with protein forming ribosomes. Each ribosome is a site where mRNA and tRNA interact, synthesizing polypeptides.

Comparing nucleic acids

Comparing nucleic acids highlights their similarities: each is a molecule of polynucleotide formed from many nucleotide units joined together by condensation reactions. Each nucleotide carries a base. However there are differences. The table summarizes the structure, make up, and function of nucleic acids.

Nucleic acids compared. Note that the base uracil (U) substitutes for thymine in mRNA and tRNA.

Questions

1 What does it mean when we say that the genetic code is non-overlapping?

2 What is a code? Use the internet to help you answer the question.

3 Explain the role of each type of RNA.

	DNA	mRNA	tRNA
Polynucleotide strand	double	single	single
Origin	replicated by parent DNA	transcribed against DNA template	transcribed against DNA template
Number of bases	variable: many thousands	variable: many hundreds to a few thousands	75–90
Pentose sugar	deoxyribose	ribose	ribose
Bases	ATCG	AUCG	AUCG
Ratio of bases	A:T = 1 C:G = 1	variable	variable
Shape of molecule	double helix	usually a single strand which can twist into a variety of shapes	single strand in the form of a 'clover leaf'
Location	mainly in the nucleus	made in the nucleus; found in the cytoplasm	made in the nucleus; found in the cytoplasm
Function	carries genetic information inherited from parent DNA	carries genetic information from the nucleus to the ribosomes where polypeptides are synthesized	binds to amino acids and carries them to the mRNA combined with ribosomes

Polypeptide synthesis is a two stage process.

- **Transcription** – the synthesis of mRNA from the DNA template of a gene. The information carried in the sequence of bases of the gene is copied in the sequence of bases of the mRNA transcribed.

- **Translation** – the conversion of the information in mRNA to make a polypeptide. The information is carried in the sequence of bases of the mRNA (itself a complementary copy of the base sequence of the gene against which the mRNA was transcribed) and determines (decides) the sequence (order) in which amino acids join together forming a polypeptide.

The diagram and its checklist are your guide to the sequence of events.

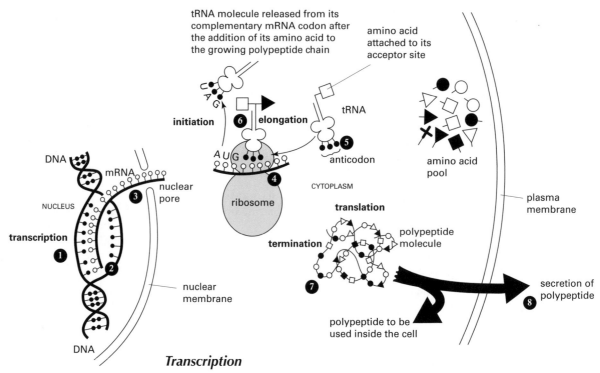

Transcription

1 • The enzyme **DNA helicase** catalyses the breaking of the hydrogen bonds that link the base pairs of a double-stranded section of DNA carrying the gene to be transcribed. The DNA unzips and its strands separate. The bases of each strand are exposed.

- Some base sequences of one of the unzipped strands (the transcribing strand) carry the genetic information which encodes polypeptide synthesis. Its unzipped partner strand does not carry genetic information.

- Each sequence of bases of the transcribing strand which encodes the synthesis of polypeptide is called an **exon**.

- The sequences of bases that do *not* encode for polypeptide are called **introns**. Sometimes introns are called 'silent' or 'junk' DNA (bacterial DNA does not contain introns).

2 • mRNA is synthesized on the transcribing DNA strand. The enzyme **RNA polymerase** first binds to a **promoter** sequence on the transcribing strand, initiating transcription. The enzyme then moves along the transcribing strand, adding RNA nucleotides to the growing mRNA strand. The bases of the nucleotides are complementary to the exposed bases of the transcribing strand.

- A **cap** of G (guanine) is added to the 5′ end, and a **poly-A tail** (150–200 adenines) to the 3′ end of the pre-mRNA molecule.

- The growing mRNA strand lengthens in the 5′ → 3′ direction.

- A strand of **pre-mRNA** (precursor mRNA) is produced. Pre-mRNA carries exons and introns which are complements of the exons and introns of the transcribing strand.

- In eukaryotic cells, further processing occurs which adds signalling sequences to and removes introns from the pre-mRNA. The process is called **editing**. Enzymes called **spliceosomes** catalyse the joining together of the exons. The result is a strand of **mature mRNA**.

3 • Strands of mature mRNA pass from the nucleus through the nuclear pores of the nuclear membrane into the cytoplasm of the cell. Its poly-A tail facilitates (makes easy) the passage of each strand.

Translation

4 • A strand of mature mRNA binds to a ribosome. Binding is facilitated by its poly A-tail. Its 5′ cap signals the point of attachment. The ribosome moves along the strand until it reaches an **initiation codon**. This is usually AUG and signals the beginning of a gene.

5 • Molecules of tRNA collect amino acids from the 'pool' of amino acids dissolved in the cytoplasm of the cell. *Remember* that each type of amino acid is carried by its own type of tRNA. The combination of an amino acid with its particular tRNA requires energy released by the hydrolysis of ATP. The process is called **activation**. The tRNA/amino acid combinations move towards a ribosome.

6 • The tRNA/amino acid combination which carries the anticodon UAC combines with its complement, the first codon AUG. This is initiation and translation begins. The second codon of the mRNA then attracts its complementary tRNA anticodon in a second tRNA/amino acid combination. The ribosome holds the two combinations in place while a peptide bond forms between the two amino acids. The reaction is catalysed by the enzyme **peptidyl transferase**.

- Once the peptide bond forms, the bond between the first (initiation) molecule of tRNA and its amino acid is hydrolysed and the unbound tRNA is released from its complementary mRNA codon. The third codon of the mRNA attracts its complementary tRNA anticodon, in a third tRNA/amino acid combination. The ribosome moves along to hold it in place while a peptide bond forms between the amino acid which its tRNA carries and the second amino acid. The bond between the second molecule of tRNA and its amino acid is hydrolysed and the unbound tRNA is released from its complementary mRNA codon.

  - As a result a strand of polypeptide forms, one amino acid at a time, according to the particular sequence of the codons of the mRNA coding tRNA/amino acids combinations to assemble in a particular order. The process is called **elongation**.

7 • The ribosome moves along the length of mRNA until it reaches a **stop codon** UAA, UGA, or UAG. This codon does not attract an anticodon but encodes a **releasing factor**. The bond between the terminal (end) tRNA molecule and its amino acid is hydrolysed and the completed polypeptide molecule released. The process is called **termination**.

- Polypeptide chains may then fold to form secondary and tertiary structures. Several chains may combine to form a quaternary structure.

- Polypeptides used inside cells are usually made on free ribosomes and released into the cytoplasm.

8 • Polypeptides to be exported from cells are usually made on the ribosomes of the endoplasmic reticulum and transported to the Golgi apparatus. Here they are modified and packaged into vesicles which bud off from the Golgi. The vesicles pass to the plasma membrane from which they are secreted by exocytosis or where they form membrane proteins.

Questions

1 Explain the difference between an exon and an intron.

2 Briefly summarize the processes of transcription and translation.

3 What is a polysome?

Fact file

siRNA interferes with gene expression

Small interfering (si)RNA is one of several types of RNA molecule that help to regulate which genes are active and how active they are. Each molecule consists of short double-stranded lengths of RNA about 20 nucleotides long.

Notice the overhang of two nucleotides at each 3′ end. siRNA is produced by the breakdown of long double-stranded RNA molecules into around 20-nucleotide lengths. An enzyme called *dicer* catalyses the breakdown.

siRNA is one of the causes of **RNA interference**. The process prevents translation of mRNA and therefore 'silences' genes.

- siRNA combines with different proteins forming a complex called **RISC** (RNA Induced Silencing Complex).
- The RISC complex scans the mRNA content of the cell in question.
- siRNA unwinds and one strand (called the **guide strand**) binds with its complementary length of mRNA.
- The binding of siRNA with its complementary length of mRNA causes the mRNA to break. Cleavage is catalysed by the enzyme *argonaute*, which is one of the RISC proteins.
- As a result translation is prevented... the gene is silenced!

Artificial siRNAs are tailor-made to be complementary to the mRNA of different genes. This means that siRNA can be used to silence specific genes. For example, silencing an oncogene would switch off the over expression of the polypeptide stimulating mitosis. In theory the rate of mitosis would slow, perhaps stopping the development of a tumour. Our ability to selectively silence genes promises exciting developments in medicine and other areas of research.

2.03 Gene mutation

Qs and As

Q Why may frame-shift mutations affect the structure and therefore function of polypeptides?

A *The function of polypeptides depends on their structure and therefore shape. Their shape is determined by hydrogen bonding and the interactions between the different R groups of the amino acids making up the polypeptide in question. The change in the sequence of R groups of the amino acid units downstream of the point mutation changes the interactions between them and the shape of the polypeptide molecule. The changed shape affects the function of the polypeptide.*

Recall that the term **mutation** refers to a change in the arrangement or the amount of genetic material in a cell. The change may affect just one gene or a chromosome or part of a chromosome carrying many genes. Mutations in sex cells (sperms or eggs) are **inherited**; we say that the mutations are **hereditary**. Mutations in other (body) cells are *not*; we say that the mutations are **acquired**.

Gene mutations are the result of copying errors in the sequence of bases of their DNA. They occur during DNA replication. If only one base is involved, then the copying error is called a **point mutation**. There are two types:

- base pair **deletions** (bases are lost) or **insertions** (bases are added)
- base pair **substitutions** (bases are replaced with others)

Deletions or insertions alter the sequence of the bases downstream of the **locus** (position of a gene on a chromosome) on the DNA strands where the mutation takes place. *Recall* that during transcription the sequence of the bases of the DNA of a gene is copied in the sequence of bases of the mRNA transcribed; and that during translation the base sequence of the transcribed mRNA determines the order in which amino acid units join up forming a polypeptide. *Recall* also that the information needed to assemble one amino acid unit in its correct place in a polypeptide is contained in the sequence of bases called a codon.

The diagram illustrates the point. *Notice:*

- The codons are set out for mRNA (U instead of T) and are therefore complements of the DNA codons from which the mRNA is transcribed. The accepted abbreviations of the amino acid encoded by the codons are included.
- The sequence of amino acids downstream of the point mutation causing either deletion or insertion is changed compared with the normal sequence encoded by the non-mutated gene. We say that the mutation causes a **frame-shift**, and frame shift mutations can significantly affect the structure and therefore function of polypeptides.

Substitutions also alter the sequence of bases within a codon, but not downstream of the mutation. The alteration may change the amino acid unit at the point of the mutation... or *not*. *Remember* that the genetic code is **degenerate**. All amino acids but two are encoded by more than one codon – in the case of the amino acid leucine there are six alternatives! This means that if the substitution replaces a base within a codon with a base of an alternative codon encoding the same amino acid, then the amino acid sequence of the polypeptide as a whole will not change, even though the base sequence of the codon has. This sort of mutation is called a **silent** mutation.

Causes of mutations

Gene mutations occur spontaneously as random events during DNA replication. However various factors in the environment increase the mutation rate. An environmental factor that causes mutation is called a **mutagen**. Ionizing radiation (e.g. X-rays, gamma rays) and different chemicals are mutagens.

- Ionizing radiation damages DNA by stripping electrons from the atoms of its molecules (ionizes the atoms).
- Alkylating agents are chemicals which transfer $-CH_3$ (methyl) and $-CH_3CH_2$ (ethyl) groups to DNA molecules, altering their activity.

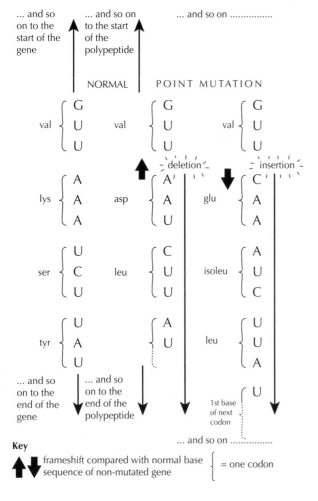

Key

⬆⬇ frameshift compared with normal base sequence of non-mutated gene

{ = one codon

Genes and cell division

Mutagens often cause mutations in the genes which control the rate of cell division. Normally cell division stops when the particular task requiring more cells is complete – a cut is healed, for example. However if mutagens cause mutations in the genes controlling cell division, then the cells proliferate and cell division runs out of control. A mass of cells called a **tumour** develops. If the cells of a tumour do not spread from the point of origin then the tumour is said to be **benign**. If they break away from the tumour and spread elsewhere in the body (**metastasis**) then the tumour is said to be **malignant**. The word **cancer** refers to malignant tumours. Mutagens which cause cancer are called **carcinogens**.

They cause mutations in:

- **proto-oncogenes** which encode proteins that stimulate normal cell division. Mutated proto-oncogenes are called **oncogenes**. Their activity results in
  - increased production of the proteins stimulating cell division or …
  - an increase in the activity of the proteins themselves
  - As a result cell division is over-stimulated and cells proliferate. A tumour develops.
- **tumour suppressor genes** which encode proteins that
  - inhibit cell division
  - attach cells to one another and anchor them in their proper place
  - repair damaged DNA before it can be replicated.

Mutation of tumour-suppressor genes inactivates them.

- As a result cell division continues when it should stop and cells proliferate. A tumour develops.

If the tumours caused because of the activity of oncogenes or mutated tumour suppressor genes undergo metastasis, then cancers develop.

The effect of mutations on the individual

We think of mutations as harmful to the individual, causing conditions such as cystic fibrosis and Down's syndrome. Many mutations like these are indeed harmful, in that they result in a faulty version of the protein the gene codes for.

However, some mutations are neutral – they do not change protein structure at all if they occur in a non-coding section of DNA, or if the changed sequence codes for the same amino acid, or for another amino acid that does not have a significant effect on the structure and function of the protein. The mutation that produced blue eyes instead of brown 6000–10 000 years ago is neutral – the colour has no effect on the function of the eye or the ability of the individual to survive.

Some mutations are beneficial, providing some of the variation that enables the species to adapt to changing conditions. For example, the mutation that produced the melanic (dark) form of the peppered moth allowed it to thrive camouflaged on dark tree bark when the light form was less well adapted. Mutations provide some of the variation that is the basis for evolution by natural selection.

Protein structure and cyclic AMP

The three-dimension structure of a protein is central to its functioning, as we have seen in considering mutations, whose effects are often caused by a change to an enzyme's active site. It is not only mutations that affect the three-dimensional structure of a protein. Extremes of pH or temperature can denature a normal protein. In addition, after their synthesis proteins are processed and packaged in the Golgi body before being released from the cell. Proteins may be produced in a deactivated form, to be activated at their site of action. In the case of adrenaline this activation is carried out by the second messenger cyclic AMP, which activates proteins by altering their three-dimensional structure.

You do not need to remember the base sequence of codons or the amino acid each codon specifies. However you may need to interpret them (or similar) in an exam.

Questions

1 What is a point mutation?

2 Explain why a mutation may be silent.

3 Why do oncogenes and mutations in tumour suppressor genes increase the risk of cancer?

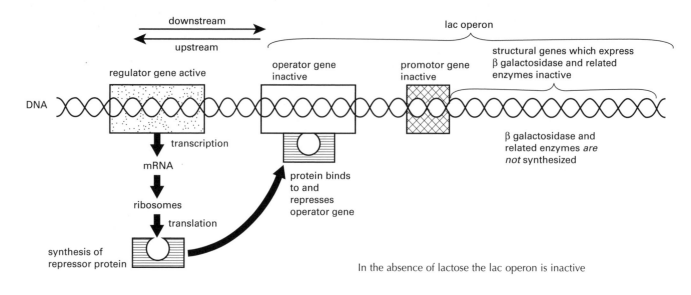

In the absence of lactose the lac operon is inactive

In bacteria groups of genes encoding proteins with related functions are arranged in units called **operons**. The operon model of gene expression was developed in 1961 by the French biologists Francois Jacob and Jacques Monod. Understanding how the **operon model** works will help you to understand how gene expression is regulated.

An operon is a length of bacterial DNA consisting of the

- **operator** which binds a protein called **repressor**
- **promoter** which binds RNA polymerase
- **structural** genes which transcribe mRNA

Another gene called the **regulator** is not part of the operon and is usually upstream of it. The regulator gene encodes repressor protein.

There are two possibilities.

- If repressor protein *binds* to the operator then RNA polymerase is not able to bind to the promoter.
   - As a result mRNA is not transcribed by the structural genes.
   - As a result the polypeptide products of the operon are not synthesized.
- If repressor protein *does not bind* to the operator then RNA polymerase is able to bind to the promoter.
   - As a result mRNA is transcribed by the structural genes.
   - As a result the polypeptide products of the operon are synthesized.

Regulating gene expression: the operon model

Transcription factors are proteins which regulate gene expression. Repressor proteins are examples. Regulation is at the level of transcription. Operons in the bacterium *Escherichia coli* illustrate the idea.

The **lac** (lactose) **operon** encodes synthesis of **beta (ß) galactosidase** and related enzymes. The enzymes catalyse the hydrolysis of lactose to glucose and galactose making the sugars available for *E. coli* to respire.

In the absence of lactose the lac operon is inactivated.

- Its regulator gene is activated and expresses repressor protein which binds to the operator.
   - As a result the operator is inactive and mRNA is unable to bind to the promoter.
   - As a result mRNA is not transcribed by the structural genes
   - As a result β galactosidase and related enzymes are not synthesized.
- The functional advantage to *E. coli* is that the bacterium conserves energy by not synthesizing enzymes unnecessarily in the absence of their substrates (lactose, glucose, and galactose).

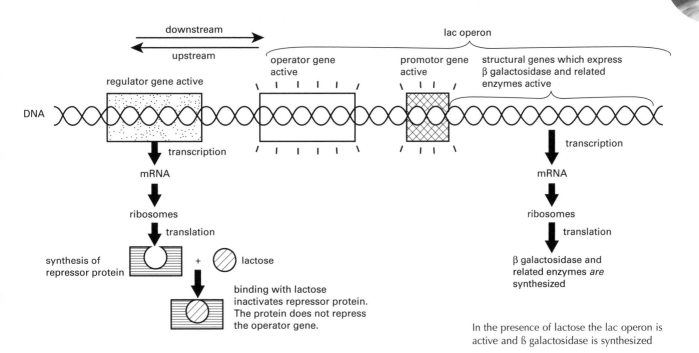

In the presence of lactose the lac operon is active and ß galactosidase is synthesized

In the presence of lactose the lac operon is activated.

- Its regulator gene is activated and expresses repressor protein to which lactose binds, inactivating it.
  - As a result the operator is active and mRNA is able to bind to the promoter.
  - As a result mRNA is transcribed by structural genes.
  - As a result β galactosidase and related enzymes are synthesized.
- The functional advantage to *E. coli* is that the enzymes which control its metabolism are available to the bacterium.

Remember that bacteria are prokaryotes. Mechanisms similar to the operon model regulating gene expression at the level of transcription occur in the cells of eukaryotes. However, the barrier of a membrane surrounding the nucleus enables regulation to take place in other ways as well.

Gene expression in eukaryotes

The genes that control development of body plans are similar in plants, animals, and fungi. The genes that control the development of an embryo all contain the same sequence of 180 nucleotides. This sequence is called the **homeobox** and it controls which sections of DNA can be transcribed.

Embryo development: homeobox genes and apoptosis

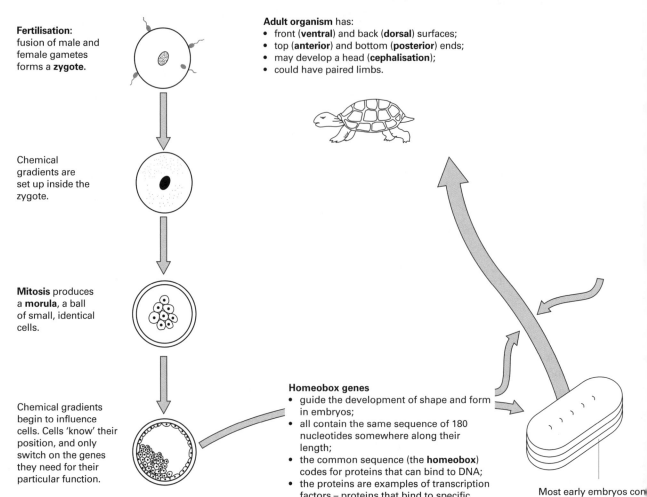

Fertilisation: fusion of male and female gametes forms a **zygote**.

Chemical gradients are set up inside the zygote.

Mitosis produces a **morula**, a ball of small, identical cells.

Chemical gradients begin to influence cells. Cells 'know' their position, and only switch on the genes they need for their particular function.

Not all of the genome is 'read' in any one cell type.

Adult organism has:
- front (**ventral**) and back (**dorsal**) surfaces;
- top (**anterior**) and bottom (**posterior**) ends;
- may develop a head (**cephalisation**);
- could have paired limbs.

Homeobox genes
- guide the development of shape and form in embryos;
- all contain the same sequence of 180 nucleotides somewhere along their length;
- the common sequence (the **homeobox**) codes for proteins that can bind to DNA;
- the proteins are examples of transcription factors – proteins that bind to specific base sequences of DNA. They control which genes are transcribed and therefore which proteins are produced by a cell.

Most early embryos con of three germ layers (**ec** **meso-** and **endoderm**) which provide the basis all tissues and organs.

Apoptosis

A form of **programmed cell death** in multicellular organisms:

- This is significant in both plant and animal tissue development.
- Tissue development involves

  cell
  ↓ mitosis
  ball of cells
  ↓ differentiation
  tissue with correct **function**
  ↓ 'pruning' by apoptosis
  tissue with correct **function** and **shape**.

- Selective apoptosis is controlled by gradients of signalling molecules – presence allows cells to multiply, absence allows them to be destroyed by apoptosis.
- Can remove useless structures, e.g. tadpole tail at metamorphosis.
- Can 'trim' structures to shape, e.g. 'paddle' limbs to hands and feet.

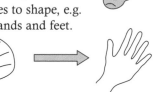

How does it happen?

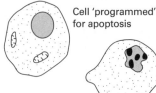

Cell 'programmed' for apoptosis

Nucleus disintegrates, 'blebs' collected by phagocyte

Nucleus condenses and cell shrinks

'blebs'

Too much or too little?

Too **much** apoptosis → tissue failure e.g. ischaemic heart disease.

Too little apoptosis → cancers.

It goes on and on!

- Between 50 and 70 million cells per day die due to apoptosis.
- In a teenager, the entire body mass is 'turned over' in a year.

2.05 Meiosis

Cell division by meiosis gives rises to gametes (sex cells). It progresses through the same phases as mitosis (with some differences), but the phases occur twice over.

- The first meiotic division is a **reduction** division which results in two daughter cells, each with half the number of chromosomes of the nucleus of the parent cell. The cells are haploid.
- The second meiotic division is a mitosis during which the two haploid daughter cells (resulting from the first meiotic division) divide.

Daughter cells which are haploid rather than diploid are an important difference which distinguishes meiosis from mitosis.

Fact file

Only the cells that give rise to gametes (the sex cells) divide by **meiosis**. Sex cells are produced in the sex organs:

- the **testes** of the male and ovaries of the female in mammals
- the **anthers** (male) and the **carpels** (female) in flowering plants.

Stages of meiosis

The detailed stages of meiosis are similar to those of mitosis that you studied at AS:

Prophase I: the chromosomes condense and are visible as two chromatids. Homologous chromosomes pair up.

Metaphase I: the spindle forms and bivalents line up on the equator; crossing over occurs.

Anaphase I: one chromosome from each bivalent moves to each end of the cell.

Telophase I: nuclear envelope reforms and cells start to divide – full cytokinesis may follow.

Metaphase II: new spindles form in each cell that resulted from the first meiotic division.

Anaphase II: centromeres divide and chromatids move to opposite ends of the cells.

Telophase II: nuclear envelopes reform and cytokinesis results in four haploid daughter cells.

Remember that, strictly, meiosis (and mitosis) refers to the processes which lead to the division of the nucleus of the parent cell. **Cytokinesis** which follows meiosis (and mitosis) describes division of the cell itself.

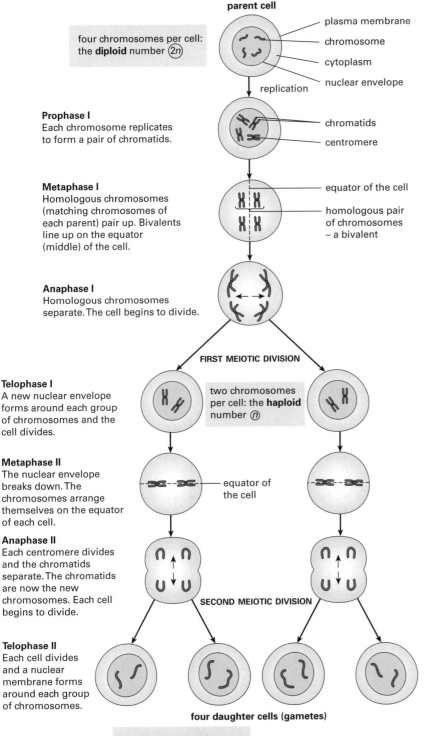

parent cell

four chromosomes per cell: the **diploid** number $2n$

plasma membrane
chromosome
cytoplasm
nuclear envelope

replication

Prophase I
Each chromosome replicates to form a pair of chromatids.

chromatids
centromere

Metaphase I
Homologous chromosomes (matching chromosomes of each parent) pair up. Bivalents line up on the equator (middle) of the cell.

equator of the cell
homologous pair of chromosomes – a bivalent

Anaphase I
Homologous chromosomes separate. The cell begins to divide.

FIRST MEIOTIC DIVISION

Telophase I
A new nuclear envelope forms around each group of chromosomes and the cell divides.

two chromosomes per cell: the **haploid** number n

Metaphase II
The nuclear envelope breaks down. The chromosomes arrange themselves on the equator of each cell.

equator of the cell

Anaphase II
Each centromere divides and the chromatids separate. The chromatids are now the new chromosomes. Each cell begins to divide.

SECOND MEIOTIC DIVISION

Telophase II
Each cell divides and a nuclear membrane forms around each group of chromosomes.

four daughter cells (gametes)

two chromosomes per cell: the **haploid** number n

Meiosis occurs in two stages. The resulting gametes have half the number of chromosomes of the parent cell.

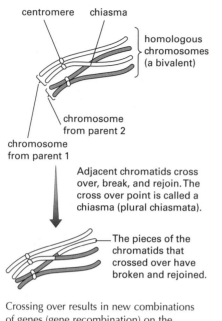

centromere   chiasma

homologous chromosomes (a bivalent)

chromosome from parent 2

chromosome from parent 1

Adjacent chromatids cross over, break, and rejoin. The cross over point is called a chiasma (plural chiasmata).

The pieces of the chromatids that crossed over have broken and rejoined.

Crossing over results in new combinations of genes (gene recombination) on the chromatids

The importance of meiosis

- Daughter cells each receive a half (**haploid** or *n*) set of chromosomes from the parent cell.
  - As a result, during **fertilization** (when sperm and egg join together) the chromosomes of each cell combine.
  - As a result, the **zygote** (fertilized egg) receives a full (**diploid** or *2n*) set of chromosomes, but inherits a new combination of the genes carried on the chromosomes (50:50) from the parents.
  - As a result, the new individual inherits characteristics from both parents, not just from one parent as in asexual reproduction.

What is the result of meiosis?

- The diploid number of chromosomes has been halved. In the diagrams, the diploid number 4 is halved to 2.
- Genetic material is exchanged between homologous chromosomes (as a result of crossing over).
- Chromosomes separate randomly.
  - As a result alleles (genes) are separated randomly – a process called **independent assortment**.
  - As a result male (paternal) and female (maternal) chromosomes are distributed randomly among the daughter cells.

Meiosis generates genetic variation

The diagram shows two pairs of chromosomes of a **sperm mother cell** (a cell which gives rise to sperm). Each chromosome is represented as a **bivalent** along which are located the genes AA, aa, BB, bb, CC, cc, DD, dd.

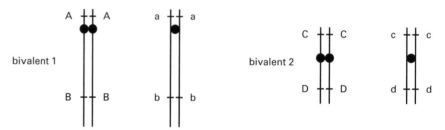

If there is only *one* crossover in each bivalent, then the possible genotypes of the sperm produced by the sperm mother cell can be mapped. The diagram shows the possible combinations of genes.

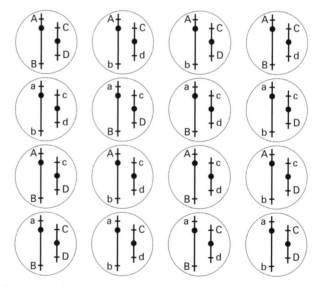

These two bivalents with one crossover can produce 16 combinations

Human cells each (except red blood cells and gametes) have 23 pairs of chromosomes – the equivalent of 92 bivalents. In each bivalent more than one crossover can occur, potentially generating an enormous variety of genetic combinations.

Questions

1 In what tissues does meiosis take place in mammals and flowering plants?

2 Explain the importance of meiosis.

3 What is the result of meiosis?

Genetics refers to the ways offspring inherit characteristics from their parents. The diagram shows how alleles controlling height are inherited when homozygous tall and short pea plants are crossed.

Some rules of genetics

- Paired genes controlling a particular characteristic are called alleles.
- Letters are used to symbolize alleles.
- A capital letter is used to symbolize the dominant member of a pair of alleles.
- The same letter in small print is used to symbolize the recessive member of a pair of alleles.

T symbolizes the allele that controls tall, and **t** symbolizes the allele that controls short.

Notice that the contrasting characteristic tall/short plants separates in the F_2 generation in a ratio of 3:1. Other characteristics of pea plants (e.g. flower colour) also separate in the F_2 generation in a ratio of approx. 3:1. The outcome of the cross allows us to state that:

- In general, when two pure-breeding individuals showing a pair of contrasting characteristics are crossed, the characteristics segregate (separate) in definite proportions in the second filial generation (F2).

Or…

- Of a pair of alleles, only one is present in a gamete.

Notice that the outcome may be explained (as the alternative statement shows) in terms of what we know of the segregation of chromosomes at meiosis. Parental alleles separate during the formation of sex cells.

Key words

Alleles – a pair of genes at the same position (**locus**) on homologous chromosomes

Homozygote – the alleles at a specific locus controlling a particular characteristic are identical

Heterozygote – the alleles at a specific locus controlling a particular characteristic are different

Dominant – any characteristic controlled by an allele that appears in preference to the form of the characteristic controlled by the allele's partner or the form of a characteristic that appears in the heterozygote

Recessive – any characteristic controlled by an allele that does not appear because the allele's partner is dominant, or any characteristic controlled by an allele that appears only in the absence of the allele's dominant partner

Genotype – the genetic make-up (all of the genes) of an individual

Phenotype – the appearance and characteristics of the cells (e.g. metabolism) of an individual resulting from those genes that are active

In the tall parent plant, both alleles which control the development of height are the same (homozygous). The parent is therefore pure breeding and produces only one kind of gamete. Every gamete carries the allele **T** which controls tallness.

In the short parent plant, both alleles which control the development of height are the same (homozygous). The parent is therefore pure breeding and produces only one kind of gamete. Every gamete carries the allele **t** which controls shortness.

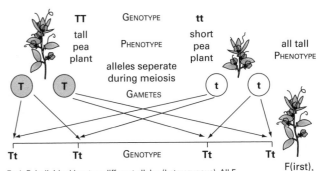

Each F_1 individual has two different alleles (heterozygous). All F_1 plants are tall, however, because **T** is dominant, masking the effect of the recessive **t**. Each F_1 plant produces two types of gamete. 50% of gametes carry the **T** allele; the other 50% carry the **t** allele.

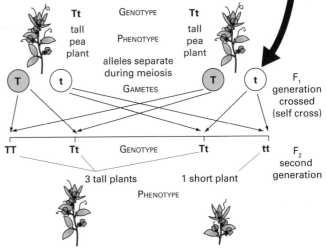

Not all the tall plants have the same combination of alleles. 50% of the plants have both dominant and recessive alleles (**Tt**) (heterozygous), and 25% are pure-breeding tall (**TT**) (homozygous). The remaining 25% are pure-breeding short (**tt**) (homozygous).

Codominance

Pea plants are either *tall* or *short*. However, some characteristics in the heterozygote are intermediate between the characteristics of the parents. For example the flowers of snap dragons may be *red* or *white*, but some may be *pink* because the alleles controlling *red* and *white* are equally dominant. The alleles are **codominant**.

The diagram on the next page sets out the possibilities.

Notice that the letters used to symbolize the alleles controlling red (R) and white (W) are capitals. A capital letter is used to denote each allele of a codominant partnership.

Multiple alleles

Sometimes more than two alleles (**multiple alleles**) encode a characteristic. An individual inherits two of the alleles available. For example, the most common type of human blood group system is controlled by three alleles, A, B and O. The A and B alleles are dominant to the O allele but not to each other (they are codominant). The table sets out the possibilities. Symbols for the alleles are I^A, I^B and I^O.

Notice that if both alleles I^A and I^B are present, then the person's blood group is AB (an example of codominance). If neither allele is present, then the person's blood group is O.

In the red-flowered parent, both alleles controlling flower colour are the same (homozygous). The parent is therefore pure breeding and produces only one kind of gamete. Every gamete carries the allele **R** for redness.

In the white-flowered parent, both alleles controlling flower colour are the same (homozygous). The parent is therefore pure breeding and produces only one kind of gamete. Every gamete carries the allele **W** for whiteness.

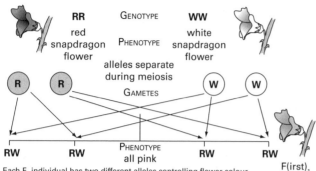

Each F_1 individual has two different alleles controlling flower colour (heterozygous). All F_1 plants are pink flowered, however, because the alleles are equally dominant. Each F_1 plant produces two types of gamete: 50% carry the **R** allele; the other 50% carry the **W** allele.

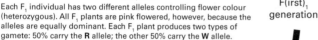

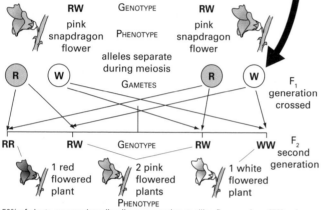

50% of plants are pure breeding (homozygous) controlling flower colour; 25% red flowered (**RR**), 25% white flowered (**WW**). The other 50% have both alleles (**RW**) and are pink flowered.

How codominant alleles controlling a characteristic (flower colour) are inherited from one generation to the next

Genotype(s)	Phenotype: antigen on surface of red blood cells	Blood group
$I^A I^A$ $I^A I^o$	I^A	A
$I^B I^B$ $I^B I^o$	I^B	B
$I^A I^B$	I^A and I^B	AB
$I^o I^o$	nil	O

Epistasis

Epistasis occurs when the allele of one gene masks or overrides the effects of another, completely separate, gene. For example, in mice one gene determines presence (**A**) or absence (**a**) of coat colour. A second gene determines solid colour (**B**) or banded colour (**b**). Gene A masks the effect of gene B, since if a mouse has genotype **aa** it will be albino (no no coat colour) and so gene B will not be expressed.

If you see an unusual ratio in a dihybrid cross that does not appear to show the normal Mendelian ratio of 9:3:3:1, such that some phenotypes are combined, this indicates the effect of epistasis. Examples include

- 15:1 ([9 + 3 +3]:1) for kernel colour in wheat
- 9:7 (9:[3 + 3 + 1]) for flower colour in sweet peas
- 12:3:1 [(9 + 3):3:1] for fruit colour in squash.

Sex-linked inheritance

Of the 23 pairs of chromosomes in the nucleus of most human cells (e.g. a skin cell), 22 pairs are similar in size and shape in both men and women. These pairs are the **autosomes** and the alleles they carry determine the phenotype of the individual other than sex (gender). The 23rd pair is the **sex chromosomes – X** and **Y**. The **X** chromosome is larger than the **Y** chromosome.

- Two X chromosomes make up the sex chromosomes of a woman. We say that she is **homogametic** because the two sex chromosomes are the same and all her gametes (eggs) will be the same, each containing an X chromosome.

- The body cells of a man carry an X chromosome and a Y chromosome. He is **heterogametic** – the two sex chromosomes are different and his gametes (sperm) will be different, each containing either an X chromosome or a Y chromosome.

The X chromosome carries genes other than those which determine sex. The characteristics which these genes control are said to be **sex-linked**. There is little space on the Y chromosome for genes, other than those that determine sex. The fact that a female is homogametic means that she may be either homozygous or heterozygous for sex-linked characteristics. This is to her advantage if the characteristic is harmful, providing the gene controlling it is recessive. In the heterozygous state the recessive gene is not expressed in the phenotype. The individual is said to be a **carrier** of the harmful recessive gene.

Because a male is heterogametic, he must be homozygous for any X-linked gene (the corresponding allele which might mask the effect of its X-linked partner is not carried on the Y chromosome). As a result the X-linked gene is expressed in the phenotype.

The gene responsible for the disease **haemophilia** is an example. The diagram shows the outcome when a woman who is a carrier of the haemophilia allele becomes a mother.

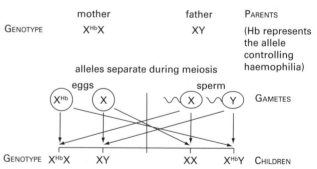

One daughter is a carrier of the haemophilia gene, one son is affected by haemophilia. The other two children are not affected by haemophilia, nor is the unaffected daughter a carrier.

Questions

1 What is the difference in meaning of the terms genes and alleles?

2 Red green colour blindness is a sex-linked condition caused by a recessive allele on the X chromosome. It occurs in 8% of men but only 0.04% of women. Explain why.

Analysis of data consisting of discrete (discontinuous) variables

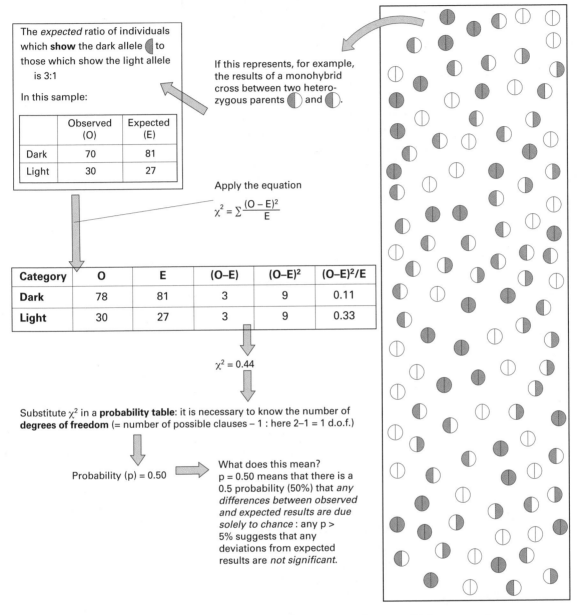

The *expected* ratio of individuals which **show** the dark allele to those which show the light allele is 3:1

In this sample:

	Observed (O)	Expected (E)
Dark	70	81
Light	30	27

If this represents, for example, the results of a monohybrid cross between two heterozygous parents and .

Apply the equation

$$\chi^2 = \sum \frac{(O - E)^2}{E}$$

Category	O	E	(O–E)	(O–E)²	(O–E)²/E
Dark	78	81	3	9	0.11
Light	30	27	3	9	0.33

$\chi^2 = 0.44$

Substitute χ^2 in a **probability table**: it is necessary to know the number of **degrees of freedom** (= number of possible clauses – 1 : here 2–1 = 1 d.o.f.)

Probability (p) = 0.50

What does this mean?
p = 0.50 means that there is a 0.5 probability (50%) that *any differences between observed and expected results are due solely to chance*: any p > 5% suggests that any deviations from expected results are *not significant*.

When performing many experiments which generate large amounts of data, statistical tests are necessary to ascertain what the results show. Applications of the Chi-squared test include to test whether certain alleles are linked. For example, if the test showed significant differences from the expected ratios of offspring when considering two phenotypes together, this may be because two genes are linked on the same chromosome, and so are not inherited independently. An example is the inheritance of flower colour and pollen grain shape in sweet peas – these features are inherited together, as they are linked on the same chromosome.

2.08 Variation and its causes

Variation and natural selection

Variation – the differences that exist between organisms either within a species or between species – is essential for the process of evolution by natural selection. Without a varied gene pool and a range of organisms adapted to different conditions, the species cannot adapt to a changed environment and cannot evolve.

Continuous variation

Some characteristics show variations spread over a range of measurements. Height is an example. All *intermediate* heights are possible between one extreme (shortness) and the other (tallness). We say that the characteristic shows **continuous variation**.

The distribution curve shows that height varies about a mean which is typical for the species. This is the same for any other continuously variable characteristic of a species.

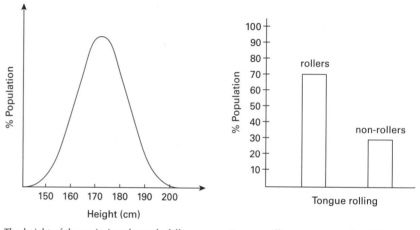

The height of the majority of people falls within the range 165–180 cm

Tongue-rolling – an example of discontinuous variation

Characteristics which vary continuously are usually the result of the activity of numerous sets of genes. We say that they are **polygenic** in origin.

Discontinuous variation

Other characteristics do not show a spread of variation. There are no intermediate forms but distinct categories. For example most human blood groups are either A, B, AB, or O; pea plants are either tall or short (dwarf). We say that the characteristics show **discontinuous variation**.

Notice in the bar chart that some people can roll the tongue, others cannot. There are no intermediate half-rollers!

Characteristics which vary discontinuously are usually the result of the activity of one set of genes. We say that they are **monogenic** in origin.

Environmental causes of variation

Variation arises from environmental causes. Here 'environmental' means all of the external influences affecting an organism. Examples are:

- **Nutrients** in the food we eat and minerals that plants absorb in solution through the roots. For example, in many countries children are now taller and heavier, age for age, than they were 50 or more years ago because of improved diet and standards of living.
- **Drugs** which may have a serious effect on appearance. For example, thalidomide was given to pregnant women to prevent them feeling sick and help them sleep. The drug can affect development of the fetus and some women prescribed thalidomide gave birth to seriously deformed children.
- **Temperature** affects the rate of enzyme-controlled chemical reactions. For example, warmth increases the rate of photosynthesis and therefore improves the rate of growth of plants.
- **Physical training** uses muscles more than normal, increasing their size and power. For example, weight-lifters develop bulging muscles as they train for their sport.

Variations that arise from environmental causes are not inherited because sex cells are not affected. Instead the characteristics are said to be **acquired**. Because the weight-lifter has developed bulging muscles does not mean that his/her children will have bulging muscles unless they take up weight-lifting as well!

Genetic causes of variation

Genetic causes and environmental causes of variation affect the structure and function of individuals. But *only* variations arising from genetic causes are inherited.

Reshuffling genes

- During meiosis, homologous chromosomes pair and then **segregate** (separate) into daughter cells following cytokinesis. The paired chromosomes segregate independently of each other. The process is called **independent assortment**.
  - As a result the sex cells (gametes) produced vary genetically, depending on the combination of chromosomes in each daughter cell.
  - As a result of random mating, parental genes are recombined in new arrangements in the **zygote** (fertilized egg).
  - As a result offspring are genetically different (vary) from each other and from their parents (except identical twins, which are genetically the same).
- Crossing over in a bivalent may involve each pair of chromatids:
  - As a result very large numbers of recombinations of chromatids (and therefore genes) are possible.
- The recombination of chromosomes (and therefore genes) following crossing over may result in major changes in the organism's genome. The activity of genes varies depending on their location on a chromosome.
  - As a result new proteins are produced.

Gene linkage

The likelihood of certain alleles to be inherited together depends on **genetic linkage**. If two loci are physically close to one another on the same chromosome then these alleles tend to stay together during meiosis, and the characteristics are genetically linked. Linked genes are less likely to be separated during crossing over.

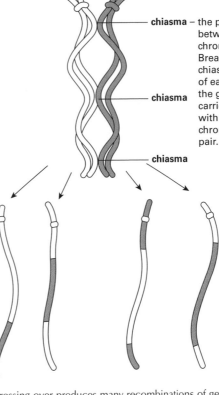

chiasma – the point of contact between the pairs of chromatids of a bivalent. Breaking occurs at the chiasma where sections of each chromatid (and the genes the section carries) are exchanged with the corresponding chromatid of the partner pair.

Crossing over produces many recombinations of genes

Differences in DNA within a gene pool

Genetic diversity is represented by the variation in the genetic material of members of a population.

- Differences between population members can be due to differences in the base sequences of the sections of DNA which form their genes. The differences arise as a result of gene mutations.
- Differences in chromosomes can arise as a result of:
  (a) crossing over and independent assortment during meiosis
  (b) the recombination of parental chromosomes in the zygote
- Whole genomes can vary as a result of differences in the base sequences of the non-coding DNA. The differences in the base sequences of non-coding DNA are also the result of mutations.

Think of a population not as a group of individuals each with a set of genes, but as a pool of genes. **Gene flow** within a **gene pool** refers to the transfer of parental genes to offspring when parents reproduce sexually.

Unrestricted gene flow maximizes genetic diversity within the gene pool. Each member of the population has the same chance as any other member of mating with an individual of the opposite sex and contributing its genes to the gene pool.

Changing genetic diversity

Restriction of gene flow reduces the size of the gene pool. In effect, the size of the population whose members can freely mate is reduced.

The frequency of genes in the gene pool of the smaller population is often different from the original larger population. The genetic diversity of small populations is therefore different from that of larger populations.

The founder effect and genetic bottleneck

The term **founder effect** refers to the change in genetic diversity in a small group of individuals compared with the genetic diversity of the larger population from which the group has separated.

The frequency of genes in the gene pool formed by the founder individuals of the small group may not be representative of the frequency of genes in the larger population from which the founder individuals have come.

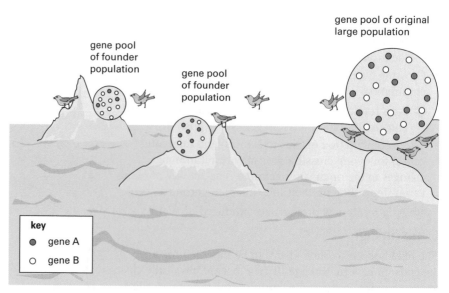

The founder effect. The frequency of genes A and B is different in the founder populations on each of the islands, and different from the original population.

The diagram illustrates how the founder effect may influence the evolution of founder populations.

- The frequency of the genes in the gene pools is different from each other and from the frequency of the genes in the gene pool of the original population.
  - As a result the effects of natural selection will cause the founder populations to diverge quickly.
- Eventually, the founder populations give rise to new species.

A **genetic bottleneck** forms when a previously large population is so reduced in numbers that only a few individuals survive.

- Like founder populations, the frequency of the genes in the gene pool of survivors is likely to be different from the frequency of genes in the gene pool of the original much larger population.
- Providing the surviving population does not become extinct, the individuals from any later increase in numbers will inherit a different frequency of genes compared with the population before the bottleneck occurred.

Research shows that founder events (whether by establishing new colonies or bottlenecks) are not likely to reduce genetic variation, unless the number of founders is very small (a single pregnant female for example). However the frequency of genes in founder populations is often different from the original population from where the founders originate.

Questions
1 What does the phrase 'genetic diversity' mean? 2 What is the difference between gene flow and a gene pool? 3 Explain the relationship between the founder effect and a genetic bottleneck.

The **Hardy–Weinberg principle** states that the frequencies of alleles and genotypes in the gene pool of a population remain constant (in equilibrium) from generation to generation unless disturbed by different influences. These include:

- selection
- non-random mating
- gene flow (migration)
- small population size
- genetic drift
- mutations

One or more 'disturbing influences' always affect natural populations so Hardy–Weinberg equilibrium is only possible in laboratory conditions. For example, in small populations chance plays an important part in determining which alleles pass from parents to offspring. The probability that the frequency of alleles will be different from one generation to the next increases, the smaller the population.

- The term **genetic drift** refers to the change in the allele frequency.
- The **founder effect** is an example of genetic drift.
- A **genetic bottleneck** may also cause genetic drift.

However the idea of genetic equilibrium is useful because it provides a standard against which change in gene frequencies in natural populations can be measured. *Remember* that the extent of change in gene frequencies of a population is a measure of its rate of evolution. So, deviation from Hardy–Weinberg equilibrium indicates the evolution of a species.

Thinking it through

Think of a population with a gene that has two alleles.

- **A** – the dominant allele: p represents its frequency in the population
- **a** – the recessive allele: q represents its frequency in the population

If we assume that all members of the population carry either of the alleles or both of them…

$$p + q = 1.0 \ (100\%) \qquad \textbf{(equation 1)}$$

… the total frequency of the alleles in the population.

Equation **1** can be used to calculate the frequency of each of the alleles in the population. For example if the frequency (p) of allele **A** in the population is 0.40 (40%) then…

$$1 - 0.40 = 0.60 \ (60\%)$$

… the frequency (q) of allele a in the population.

Remember that most organisms are diploid and therefore carry the alleles of a gene in pairs. At meiosis a proportion of gametes will carry the **A** allele (frequency p); likewise a proportion of gametes will carry the **a** allele (frequency q). On fertilization the gametes combine at random to form new genotypes as:

Possible pairings		Frequency
A	A	$p \times p = p^2$
A	a	$p \times q$
a	A	$q \times p$ $\Big\}$ = $2pq$
a	a	$q \times q = q^2$

Notice that the frequency of:

- **AA** (homozygous) dominant genotype is p^2, and ¼ of the total possible genotypes. An individual has ¼ (25%) chance of being **AA**
- **Aa** (heterozygous) genotype is $pq + pq$ is $2pq$, and ½ of the total possible genotypes. An individual has ½ (50%) chance of being **Aa**
- **aa** (homozygous) recessive genotype is q^2, and ¼ of the total possible genotypes. An individual has ¼ (25%) chance of being **aa**

Remember

- A population is made up of a group of individuals of the same species living in the same place at the same time. A species may be represented by only one population but most have more.

- An individual's genotype is the total of all the alleles of all its genes. We can think of a population as a group of genotypes and all the alleles of all of the genes represented by the genotypes. The total of all the alleles of all the genes of a population is called a **gene pool**.

Fact file

G.H. Hardy was a prominent pure mathematician at Cambridge in the early 1900s. In 1908 the geneticist R.C. Punnett, a friend of Hardy and also at Cambridge, was puzzled by the relationship between allele frequencies and phenotype frequencies. He asked Hardy for an explanation, who promptly replied "p^2 to $2pq$ to q^2". At the same time the German physician W. Weinberg came to a similar conclusion. This is why the idea is called the Hardy–Weinberg principle.

or

- homozygous dominant ¼ (25%) + heterozygous ½ (50%) + homozygous recessive ¼ (25%) = 1.0 (100% of total possible genotypes)

or

- **AA** + 2**Aa** + **aa** = 1.0 (100%)

  which may be expressed as…

- $p^2 + 2pq + q^2 = 1.0$ (100%) (**equation 2**)

  … the **Hardy–Weinberg principle**

Put in words, the Hardy-Weinberg principle states that if the frequency of one allele (**A**) is p and the frequency of the other allele is q then…

$$p + q = 1 \text{ (equation 1)}$$

….then the frequencies of the three possible genotypes are p^2 (**AA**), $2pq$ (**Aa**) and q^2 (**aa**), so that

$$p^2 + 2pq + q^2 = 1 \text{ (equation 2)}$$

… giving, therefore, the relationship between allele frequencies and genotype frequencies.

Notice that equation 2 is equation 1 to the power^2

$$(p + q)^2 = 1$$

…. because it takes into account that most organisms are diploid.

Notice also that there are only two phenotypes:

- homozygous recessive   **aa** }
- homozygous dominant   **AA** ⎫
  heterozygous   **Aa** ⎭

…. therefore the relationship between allele frequencies and phenotype frequencies is:

   frequency of the recessive phenotype = q^2

   frequency of the dominant phenotype = $p^2 + 2pq$

An afterthought

It is always best to use proportions when using the Hardy–Weinberg principle rather than percentages, e.g. $2pq = 0.32$ is better than $2pq = 32\%$.

Questions

1 What is a gene pool?

2 Under what conditions will the frequencies of alleles in a gene pool remain constant?

3 Of a population of sheep, 25% have black wool. The allele controlling white wool is dominant to the allele controlling black wool. What are the frequencies of the genotypes controlling wool colour in the population?

Working it through

Equations 1 and 2 can be used to calculate the frequency of any allele and genotype in a population. For example **sickle cell anaemia** is widespread in Africa and parts of India and the Mediterranean. It is caused by a recessive mutation of one of the alleles of one of the genes that control the synthesis of the oxygen absorbing pigment haemoglobin.

For the gene in question let…

   Hb^A symbolize the dominant allele encoding the synthesis of normal haemoglobin

   Hb^s symbolize the recessive allele encoding the synthesis of sickle haemoglobin

Asking the question…

In some parts of Africa the proportion of individuals homozygous for the recessive sickle allele is 4%. Assuming Hardy–Weinberg equilibrium, what proportion of the population would be expected to be heterozygous?

The frequency of homozygous recessives $(q^2) = 0.04$ (4% **Hb^sHb^s**)

Remember that ….

   $p + q = 1$ (equation 1) where

   p = frequency of **Hb^A**

   q = frequency of **Hb^s**

… *therefore*

   $q = \sqrt{(0.04)} = 0.2$

   $p = 1 - 0.2 = 0.8$

… *therefore* using equation 2

   $p^2 = (0.8)^2 = 0.64$ (64% **Hb^AHb^A**)

… *therefore*

   $2pq = 2 \times 0.8 \times 0.2 = 0.32$ (32% **Hb^AHb^s**)

… the proportion of the population expected to be heterozygous.

So, knowing the frequency of the recessive allele **Hb^s**(q) we have been able to calculate the frequency of the dominant allele **Hb^A** (p) using equation 1. We have also been able to calculate the frequency of the three possible genotypes by taking these answers and applying them in equation 2.

Fact file

Present-day living things are descended from ancestors that have changed through time (evolved) over thousands of generations. The lifetime's work of the English naturalist Charles Darwin (1809–1882) provided much evidence that organisms evolve. His ideas on *how* organisms evolve were even more important. He proposed a mechanism for evolution. The mechanism is **natural selection**. In 1859 the proposal was published in his book *On the Origin of Species by Means of Natural Selection*.

Qs and As

Q Evolution is sometimes defined as a result of changes in the frequency of particular alleles. Explain the definition in terms of natural selection.

A *In the case of stabilizing selection, the frequency of alleles controlling the 'average' form of the characteristic(s) in question increases, whereas the frequency of the alleles controlling the 'extreme' forms of the characteristic decreases.*

In the case of directional selection, the frequency of the alleles controlling one of the extreme forms of the characteristic increases but the frequency of alleles controlling the 'average' form and the other 'extreme' form of the characteristic(s) decreases.

Fact file

A species is:

'the lowest taxonomic group';

'a group of organisms which can interbreed and produce fertile offspring';

'a group of organisms which share the same ecological niche';

… and members of the same species have a very high proportion of their DNA in common.

Natural selection works on the genetic variation in the gene pool. Alleles which favour survival are selected for and their **frequency** (percentage occurrence in the gene pool) changes. The characteristic(s) controlled by the favourable alleles change(s) enabling individuals with the alleles to adapt to a changing environment. The population evolves. The extent of change in the frequency of an allele (or alleles) is a measure of the rate of evolution of the population in question. Change in an environment stimulating evolution represents **selection pressure**.

Trends in natural selection

In stable environments selection pressure is low. For any particular characteristic (e.g. length of body), individuals with extremes of the characteristic (bodies which are short or long) are selected *against*. Those close to the average are selected *for* and are therefore more likely to reproduce (differential reproductive success), passing on the alleles controlling 'averageness' to the next generation. If the trend is long term, the species changes very little. Descendents, therefore, look like their distant ancestors. This kind of selection maintains the constancy of a species. It is called **stabilizing selection**.

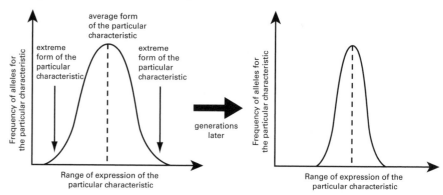

Over time, stabilizing selection reduces the frequency of the alleles for the extreme expression of a particular characteristic (e.g. length of body)

Where the environment is rapidly changing, selection pressure is high and new species quickly arise. Selection pressure promotes the adaptation of individuals to the altering circumstances.

For example, a longer body may favour survival if predators find that catching shorter bodied individuals is relatively easy. In these circumstances individuals with short bodies are selected *against* and individuals with long bodies are selected *for*. Long bodied individuals are therefore more likely to reproduce (differential reproductive success), passing on the alleles for long body to the next generation. If the trend is long term, the species changes. Descendents therefore do not look like their ancestors and become a distinct and different species. This kind of selection is called **directional selection**. Once the new characteristic is at its optimum (maximizes the survival chances of individuals) with respect to the new environment, then stabilizing selection takes over.

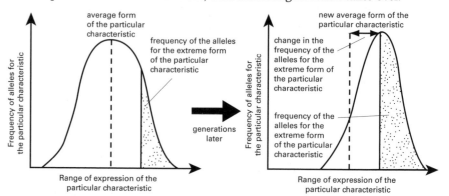

Over time directional selection increases the frequency of the alleles for an extreme expression of a characteristic (or characteristics) which become(s) the 'new average' of the evolving species

Speciation

Speciation (the formation of new species) is the result of natural selection and the outcome of evolution. It occurs when isolating mechanisms lead to divergence of gene pools. By 'isolating mechanisms' and 'divergence of gene pools' we mean that different events interrupt the free flow of alleles from parents to offspring within the gene pool of a population.

What was a single population is separated into components each called a **deme**. The individuals of each deme encounter and respond to the slightly different circumstance of the fragmented environment, so that the process of speciation gets underway. The response over time of generations of individuals of a deme is adaptation through natural selection to the slightly different circumstances. Geographic separation brings about fragmentation of a population into demes isolated from one another.

- Separation may be local (e.g. the formation of a river or mountain range or even building a road) or large scale separation of land masses (continental drift).
  - Ⓡ As a result the meeting of and matings between individuals of the original population are interrupted.
  - Ⓡ As a result the free flow of genes is restricted to individuals of each deme.
- The individuals of each deme respond to their particular environment, which may be different from the environment encountered by the individuals living in other demes.
  - Ⓡ As a result the frequency of alleles in the gene pool of each deme changes differently from the gene pool of each of the other demes.
  - Ⓡ As a result divergence occurs as the individuals of the original population adapt within their respective deme to the environmental circumstances affecting them.
  - Ⓡ As a result of divergence, over time new races, sub-species and eventually new species emerge.

The term **allopatric speciation** refers to the process. The diagram shows you the idea.

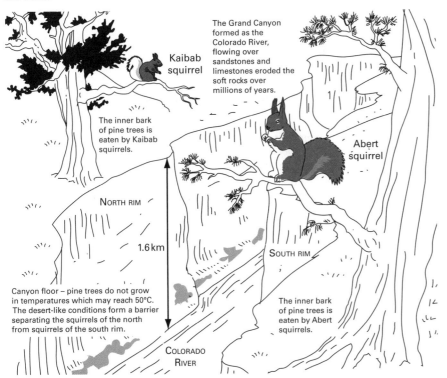

The Kaibab squirrels of the north rim of the Grand Canyon Arizona, USA and the Abert squirrel of the south rim were probably one and the same species. Prolonged isolation, as a result of the barrier of the desert-like environment of the canyon floor, has given rise to the differences in colouration between individuals of the demes we see today. The different types do not interbreed and are probably two distinct species.

Fact file

Races or sub-species are extreme variants of a particular species. Their emergence in an environment suggests fragmentation of the environment into components each with slight differences in the conditions affecting the population of the species. These differences each represent selection pressure to which each deme responds. The emergence of races or sub-species occurs at an early stage in the process.

If then the mechanisms fragmenting the original population into demes stop having an effect, then individuals of the different races or sub species will be able to interbreed, and stabilizing selection will re-establish the *status quo*.

If however the effects of the isolating mechanisms persist, then speciation will continue and new species result. If then the isolating mechanisms cease to have an effect, stabilizing selection will not re-establish the *status quo*, as individuals of the different species will not be able to interbreed or, if they do, the offspring will probably be sterile.

Mechanisms for speciation

Speciation may come about as a result of different isolating mechanisms:
- ecological (geographic) separation, where the population is separated by a physical barrier
- seasonal (temporal) separation; for example, plants may flower at different times and so be incapable of interbreeding
- reproductive separation: mating behaviours such as courtship displays may become so different that mating does not take place between the different groups, or the populations may evolve to the point where mating is not physically possible.

Questions

1. Explain the different evolutionary outcomes of stabilizing selection and directional selection.
2. Define evolution.
3. Give examples of circumstances which fragment a population into demes.

Artificial selection occurs when humans, rather than environmental factors as in natural selection, determine which genotypes will pass to successive generations.

Polyploidy and plant breeding

Polyploids contain **multiple sets of chromosomes** (chromosome multiplication can be induced by treatment with **colchicine** during mitosis – this inhibits spindle formation and prevents chromatid separation).

Autopolyploids (all chromosomes from the **same** species) e.g. all **bananas** are **triploid** – they are infertile and contain no seeds. Most **potatoes** are **tetraploid** – cells are bigger and tubers are larger. Cultivated **strawberries** are **octoploid**.

Allopolyploidy (sets of chromosomes from **different** species) is possible if the two species have a chromosome complement similar in number and shape. This might allow plant breeders to **combine the beneficial characteristics of more than one species.**

The evolution of **bread wheat** (*Triticum aestivum*) is an important example.

Wild wheat: has brittle ears which fall off on harvesting.

Selective breeding

Einkorn wheat: non-brittle but low yielding.

Polyploidy with *Agropyron* grass

Emmer wheat: high yielding but difficult to separate seed during threshing.

Polyploidy with *Aegilops* grass

Bread wheat: high yielding with easily separated 'naked' seeds.

Inbreeding and outbreeding

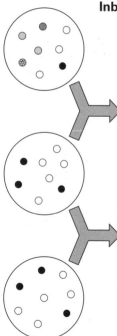

Outbreeding occurs when there is selective controlled reproduction between members of genetically distant populations (different strains or even, for plants, closely related but different species).

Tends to **introduce new and superior phenotypes**: the progeny are known as **hybrids** and the development of improved characteristics is called **heterosis** or **hybrid vigour**.

e.g. introduction of disease resistance from wild sheep to domestic strains;
combination of shorter-stemmed 'wild' wheat and heavy yielding 'domestic' wheat.

This may result from increased numbers of dominant alleles or from new opportunities for gene interaction.

Inbreeding occurs when there is selective reproduction between closely related individuals, e.g. between offspring of same litter or between parent and child.

Tends to **maintain desirable characteristics**

e.g. uniform height in maize (easier mechanical harvesting);
maximum oil content of linseed (more economical extraction);
milk production by Jersey cows (high cream content).

But it may cause **reduced fertility** and **lowered disease resistance** as genetic variation is reduced.

Thus inbreeding is not favoured by animal breeders.

Techniques with animals are less well advanced than those with plants because:

- animals have a longer generation time and few offspring;
- more food will be made available from improved plants;
- there are many ethical problems which limit genetic experiments with animals.

Two important animal techniques are:

Artificial insemination: allows sperm from a male with desirable characteristics to fertilise a number of female animals.

Embryo transplantation: allows the use of **surrogate mothers** (thus increasing number of offspring) and **cloning** (production of many identical animals with the desired characteristics).

Selection of dairy cattle
- Identify parents that produce offspring with high milk yield
- Father can now inseminate many daughters.

Desirable features:
- high yield of milk;
- high butterfat content;
- docile during milking;
- teat placement on udders.

2.13　Cloning in plants

What is cloning?

Clones are organisms that are genetically identical – that have the same genome. Cloning happens in nature in vegetative reproduction in plants, and also in some animals such as *Hydra*. In **reproductive cloning**, plants or animals are cloned artificially to produce more individuals with desirable characteristics.

In **non-reproductive cloning** (sometimes called therapeutic cloning) an embryo is created by replacing the nucleus of an egg cell with the nucleus of a cell from another individual (an embryo, fetus, or adult). Embryonic stem cells are removed from the embryo and these cells are then grown to produce tissues for transplantation.

Vegetative propagation

The natural processes of asexual reproduction are used to propagate plants in horticulture and forestry. Plants naturally produce organs of vegetative propagation, including:

- bulbs (stems with underground leaves), e.g. daffodils and onions
- corms (swollen stems), e.g. crocuses
- rhizomes (horizontal underground stems), e.g. mint, couch grass
- runners (lateral stems above the soil), e.g. strawberry
- suckers (branches of a stem), e.g. raspberry
- tubers (swollen stems or roots), e.g. potato.

English elm trees (*Ulmus procera*) are propagated from root suckers which the tree produces naturally when under stress. This helps the tree survive difficult environmental conditions as the root suckers allow a new tree to grow a few metres away. This feature can be used by horticulturists, who remove the root suckers and grow them on in controlled conditions to produce many identical trees.

Tissue culture

Another method of vegetative reproduction which is artificial, in that it does not use the plant's natural reproductive mechanisms, is **tissue culture** (micropropagation).

The technique makes use of the totipotency of meristems and the ability of differentiated plants cells to revert to a totipotent state. Growing **explants** (the part of the plant to be cultured) *in vitro* in tissue culture enables us to grow thousands of genetically identical plants (**clones**) from a single parent plant.

- Very small pieces of the plant are removed and surface sterilised. These are the explants.
- Explants are placed in growth medium (a liquid or gel) that contains all the ingredients that the pieces of tissues need to grow. This includes mineral salts, vitamins and plant growth substances.
- Conditions are sterile.
  - As a result, the new plants are free of disease.
- Temperature is optimized, encouraging growth.

Successfully growing explants in tissue culture into healthy plants depends on the

- *growth medium* – different mixtures of plant growth substances, salts and other nutrients are used, depending on the plant species and the product wanted: for example, the growth medium used to produce embryoid bodies for 'artificial seeds' is different from the medium used to produce whole plantlets
- *culture environment* – explants grow only if pH and light conditions are suitable for the plant species in question
- *growing conditions* – plantlets are more delicate than seedlings. Some form of protection (plastic sheeting, greenhouse) helps them to survive the early stages of growth following planting out into trays filled with a suitable artificial soil mix. After several weeks the plantlets are more hardy and able to survive in the open.

Advantages and disadvantages of tissue culture

The advantages of producing clones of plants by tissue culture are that the plants are

- healthy
- genetically identical to each other (clones) and to the parent plant.
  - As a result all of the desirable qualities of the parent plant (e.g. flower colour, fruit quality, resistance to disease) are retained in the clone.
- By manipulating the environmental conditions, the process can be used at any time of year so is not dependent on the plant's natural seasonal growth.

Producing vast numbers of genetically identical plants by tissue culture also has some disadvantages.

- They are all susceptible to the same diseases, so the population cannot evolve in response to a disease or unfavourable environmental change.
- The process is expensive, being labour intensive and involving the use of technology and high energy costs.

How are animals cloned?

Animals are cloned artificially in two ways.

- An embryo is produced by the fertilisation of an egg by a sperm. The embryo is then split to produce several new individuals.
  - These can be returned to the natural mother's womb.
  - Alternatively they can be placed in a **surrogate** mother's womb to develop. The surrogate mother may be of the same or a different species.
- In **somatic nuclear transfer**, the nucleus is removed from an egg and the nucleus from a body cell of another donor organism is placed into the egg. The embryo then develops in a surrogate mother into an individual that has the characteristics of the donor. Dolly the sheep was a product of nuclear transfer – the donor nucleus came from a mammary gland cell of a Finn Dorset ewe, while the egg cell came from a Scottish Blackface ewe. The result was a Finn Dorset.

Advantages and disadvantages of animal cloning

Advantages:

- A valuable animal with desirable characteristics can be used to quickly produce many identical offspring by embryo splitting.
- Cloned laboratory animals can be produced which are useful for medical research, since genetic differences can be ruled out of experiments.
- Embryos can be screened before being implanted into the surrogate, so that only healthy embryos are grown on.
- Nuclear transfer can be used to produce clones of genetically altered organisms, for example transgenic sheep that have been created to secrete human proteins in their milk.
- Nuclear transfer is also useful in creating new individuals of endangered species, since only one adult parent is needed rather than a fertile male and female.

Disadvantages:

- Many people have ethical objections to the use of embryos, and also to the process of nuclear transfer.
- There are strict rules about research into embryos and embryo stem cells.
- The success rate of nuclear transfer is very low and the process is very expensive and inefficient.

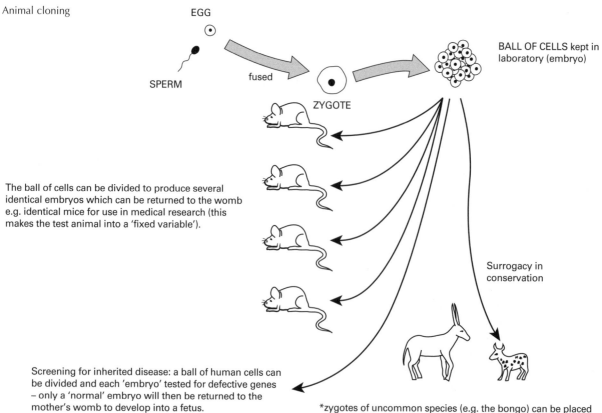

Animal cloning

EGG

SPERM

fused

ZYGOTE

BALL OF CELLS kept in laboratory (embryo)

The ball of cells can be divided to produce several identical embryos which can be returned to the womb e.g. identical mice for use in medical research (this makes the test animal into a 'fixed variable').

Surrogacy in conservation

Screening for inherited disease: a ball of human cells can be divided and each 'embryo' tested for defective genes – only a 'normal' embryo will then be returned to the mother's womb to develop into a fetus.

*zygotes of uncommon species (e.g. the bongo) can be placed in the womb of a common species (e.g. eland) – in this way the maximum number of 'rare' offspring can be produced.

2.15 Biotechnology

What is biotechnology?

Biotechnology is the industrial use of living organisms (or parts of living organisms) to produce food, drugs, or other products.

Micro-organisms such as yeast and bacteria are commonly used in biotechnological processes because they

- reproduce quickly, so can create large amounts of product in a relatively short time
- carry out reactions at moderate temperatures giving great fuel savings over non-biological production methods
- produce less waste and purer products than non-biological production methods
- can be genetically engineered to produce compounds needed for humans.

Standard growth curve of a microorganism in closed culture

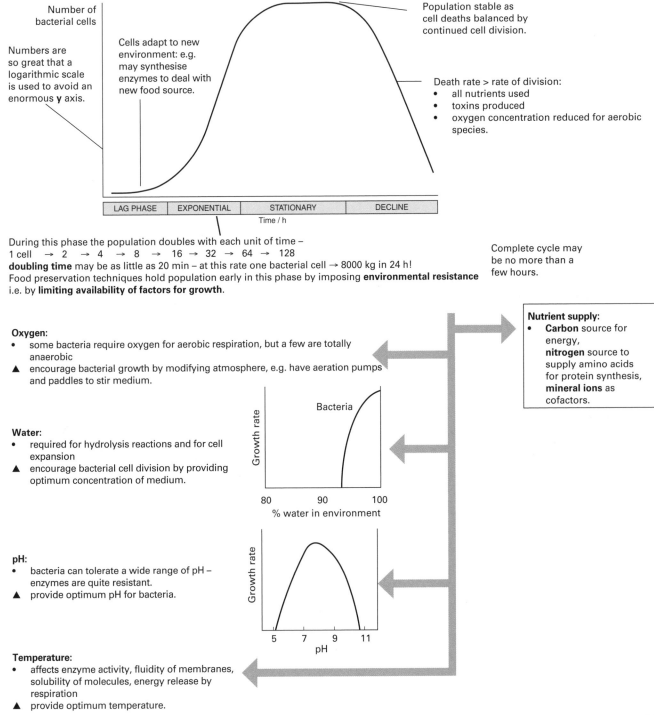

Number of bacterial cells

Numbers are so great that a logarithmic scale is used to avoid an enormous **y** axis.

Cells adapt to new environment: e.g. may synthesise enzymes to deal with new food source.

Population stable as cell deaths balanced by continued cell division.

Death rate > rate of division:
- all nutrients used
- toxins produced
- oxygen concentration reduced for aerobic species.

| LAG PHASE | EXPONENTIAL | STATIONARY | DECLINE |

Time / h

During this phase the population doubles with each unit of time –
1 cell → 2 → 4 → 8 → 16 → 32 → 64 → 128
doubling time may be as little as 20 min – at this rate one bacterial cell → 8000 kg in 24 h!
Food preservation techniques hold population early in this phase by imposing **environmental resistance** i.e. by **limiting availability of factors for growth**.

Complete cycle may be no more than a few hours.

Nutrient supply:
- **Carbon** source for energy,
  nitrogen source to supply amino acids for protein synthesis,
  mineral ions as cofactors.

Oxygen:
- some bacteria require oxygen for aerobic respiration, but a few are totally anaerobic
- ▲ encourage bacterial growth by modifying atmosphere, e.g. have aeration pumps and paddles to stir medium.

Water:
- required for hydrolysis reactions and for cell expansion
- ▲ encourage bacterial cell division by providing optimum concentration of medium.

Growth rate

Bacteria

80   90   100
% water in environment

pH:
- bacteria can tolerate a wide range of pH – enzymes are quite resistant.
- ▲ provide optimum pH for bacteria.

Growth rate

5   7   9   11
pH

Temperature:
- affects enzyme activity, fluidity of membranes, solubility of molecules, energy release by respiration
- ▲ provide optimum temperature.

Glucose oxidase

The reaction catalysed by glucose oxidase is

$$\beta-D-glucose + O_2 \longrightarrow gluconic\ acid + H_2O_2$$

The quick and accurate measurement of glucose is of great importance both medically (in sufferers from diabetes, for example) and industrially (in fermentation reactions, for example). A simple quantitative procedure can be devised by coupling the production of hydrogen peroxide to the activity of the enzyme **peroxidase**.

$$DH_2 + H_2O_2 \xrightarrow{\text{peroxidase}} 2H_2O + D$$

chromagen   coloured
a hydrogen donor   compound
(colourless)   (colour)

Peroxidase can oxidise an organic chromagen (DH_2) to a coloured compound (D) utilising the hydrogen peroxide – the amount of the coloured compound D produced is a direct measure of the amount of glucose which has reacted. It can be measured quantitatively using a colorimeter or, more subjectively, by comparison with a colour reference card.

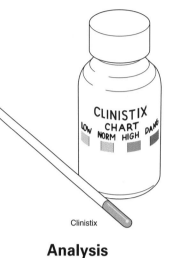

Clinistix

Analysis

Commercial applications of enzymes

Medicine

This method of glucose analysis is **highly specific** and has the enormous advantage over chemical methods in that this specificity allows glucose to be assayed **in the presence of other sugars**, e.g. in a biological fluid such as blood or urine, without the need for an initial separation.

Both of the enzymes glucose oxidase and peroxidase, and the chromagen DH_2, can be immobilised on a cellulose fibre pad. This forms the basis of the glucose dipsticks ('Clinistix') which were developed to enable diabetics to monitor their own blood or urine glucose levels.

Textiles
Subtilisin is a bacterial protease (protein → amino acids) which is used in bioactive detergents to remove protein stains from clothes.

Food production
Amylase and **glucose isomerase** convert starch → high fructose syrups, these syrups have enhanced sweetening power and lowered energy content.

There are many applications of enzyme technology to industry. Enzyme technology has several advantages over 'whole-organism' technology.

1. **No loss of substrate due to increased biomass.** For example, when whole yeast is used to ferment sugar to alcohol it always 'wastes' some of the sugar by converting it into cell wall material and protoplasm for its own growth.
2. **Elimination of wasteful side reactions.** Whole organisms may convert some of the substrate into irrelevant compounds or even contain enzymes for degrading the desired product into something else.
3. **Optimum conditions for a particular enzyme may be used.** These conditions may not be optimal for the whole organism – in some organisms particular enzymes might be working at less than maximum efficiency.
4. **Purification of the product is easier.** This is especially true using immobilised enzymes.

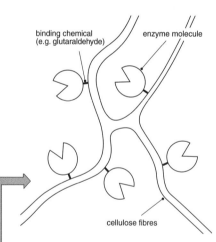

binding chemical (e.g. glutaraldehyde) enzyme molecule

cellulose fibres

Enzyme immobilisation
Immobilisation means physically or chemically trapping enzymes or cells onto surfaces or inside fibres. The benefits can be considerable:
- the same enzyme molecules can be used again and again, since they are not lost;
- the enzyme does not contaminate the end product;
- the enzymes may be considerably more stable in immobilised form – for example, glucose isomerase is stable at 65 °C when immobilised.

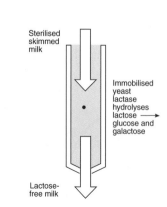

Sterilised skimmed milk

Immobilised yeast lactase hydrolyses lactose ⟶ glucose and galactose

Lactose-free milk

An important medical application of an immobilised enzyme

Some adults are **lactose-intolerant** since they lack an intestinal lactase, and undigested lactose in the gut is metabolised by bacteria causing severe abdominal pain and diarrhoea.

Milk is an important dietary component and can be made **lactose-free** by passage down a column packed with **yeast lactase** immobilised on fibres of cellulose acetate.

How are enzymes immobilised?

Cross linkage

Cross linking can damage some enzymes but enzymes that are not damaged remain very active.

enzyme — CH — (CH₂)₃ — CH = CH — (CH₂)₃ — CH=

glutaraldehyde

enzyme

covalent bond to cross-linking agent

Entrapment

enzyme cannot be washed out

the rate at which the substrate diffuses in may slow down the reaction

gel micro-capsule

fibrous polymer mesh

Entrapment is the most gentle method of immobilisation and does not damage enzymes

Adsorption

enzyme held by weak forces and may become detached

absorbing agent such as glass bead, carbon particle or collagen

Adsorption makes it easy for the enzyme to come into contact with its substrate but the process is expensive

Good news or bad?

Good:
Enzymes are not mixed with product, so recovery costs are low.
Enzymes are more stable because conditions can be controlled close to the enzyme.
Process is continuous – enzyme is re-used and reactor only requires occasional cleaning.

Bad:
Equipment is expensive, and control systems are very complex.
Some immobilised enzymes are less active.
Contamination is very expensive because the complete reactor must be closed down.
Substrate and product may block the column.

Commercial production of enzymes

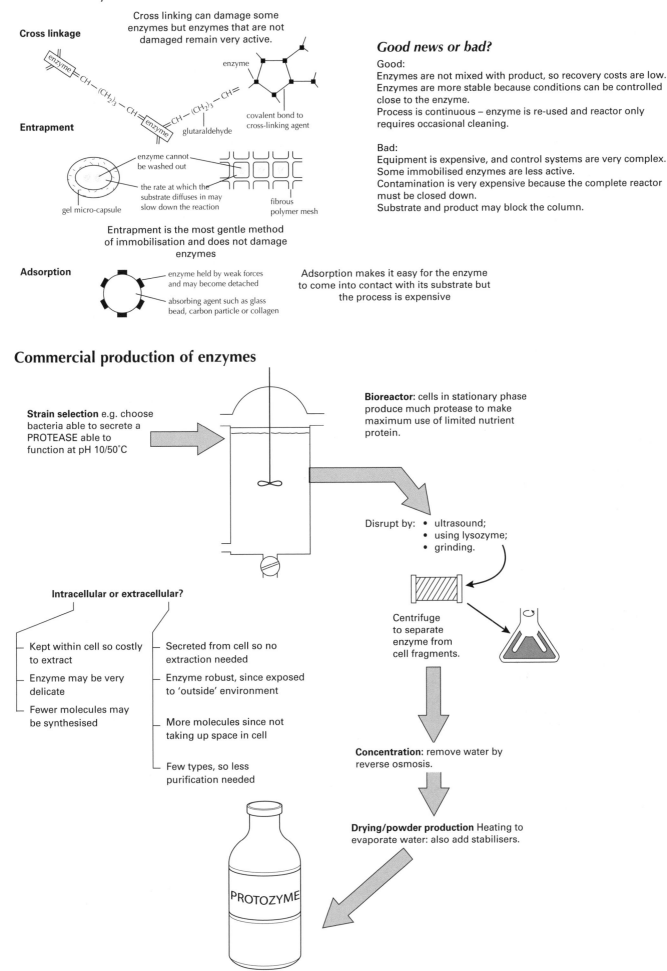

Strain selection e.g. choose bacteria able to secrete a PROTEASE able to function at pH 10/50°C

Bioreactor: cells in stationary phase produce much protease to make maximum use of limited nutrient protein.

Disrupt by:
• ultrasound;
• using lysozyme;
• grinding.

Centrifuge to separate enzyme from cell fragments.

Intracellular or extracellular?

— Kept within cell so costly to extract

— Enzyme may be very delicate

— Fewer molecules may be synthesised

— Secreted from cell so no extraction needed

— Enzyme robust, since exposed to 'outside' environment

— More molecules since not taking up space in cell

— Few types, so less purification needed

Concentration: remove water by reverse osmosis.

Drying/powder production Heating to evaporate water: also add stabilisers.

PROTOZYME

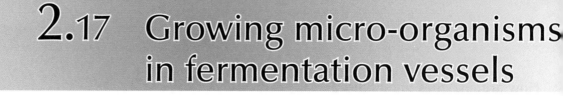

Bioreactors/fermenters exploit microbes for commercial reasons.

Paddle stirrers: continuously mix the contents of the bioreactor
- ensure micro-organisms are always in contact with nutrients
- ensures an even temperature throughout the fermentation mixture
- for aerobic (oxygen-requiring) fermentations the mixing may be carried out by an **airstream**.

Gas outlet: gas may be evolved during fermentation. This must be released to avoid pressure damage, and may be a valuable by-product e.g. carbon dioxide is collected and sold for use in fizzy drinks.

Input for microbial culture: the organisms which will carry out the fermentation process are cultured separately until they are growing well.

Nutrient input: the micro-organisms require:
- an **energy source** – usually carbohydrate
- **growth materials** – amino acids, or ammonium salts which can be converted to amino acids, are required for protein synthesis.

Probes: monitor conditions such as pH, temperature and oxygen concentration. Send information to computer control systems which correct any changes to maintain the optimum conditions for fermentation.

HEATING/ COOLING WATER OUT

Constant temperature water jacket: the temperature is controlled so that it is high enough to promote enzyme activity but not so high that enzymes and other proteins in the microbes are denatured.
- This initially involves **heating** to initiate the fermentation.
- Fermentation releases heat so that later stages may require **cooling**.

HEATING/ COOLING WATER IN

Sterile conditions are essential: the culture must be pure and all nutrients/equipment sterile to:
- avoid competition for expensive nutrients
- limit the danger of disease-causing organisms contaminating the product.

Vessel walls: made of stainless steel
- does not corrode as is unaffected by fermentation products
- can be easily cleaned and sterilised using steam jets.

Waste not, want not!
- Ideally the nutrients should be waste products of other reactions e.g. remains of crops grown for other purposes.
- The micro-organisms should be used to start or maintain new cultures.

Further processing of product may be necessary:
- to separate the micro-organism from the desired product. In some fermentation systems these micro-organisms may then be returned to the vessel to continue the process.
- to prepare the product for sale or distribution – this often involves **drying** or **crystallisation**.

'Batch' or 'continuous' culture

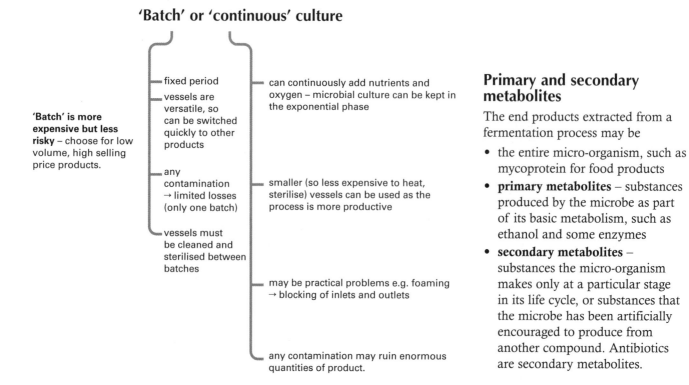

'Batch' is more expensive but less risky – choose for low volume, high selling price products.

- fixed period
- vessels are versatile, so can be switched quickly to other products
- any contamination → limited losses (only one batch)
- vessels must be cleaned and sterilised between batches

- can continuously add nutrients and oxygen – microbial culture can be kept in the exponential phase
- smaller (so less expensive to heat, sterilise) vessels can be used as the process is more productive
- may be practical problems e.g. foaming → blocking of inlets and outlets
- any contamination may ruin enormous quantities of product.

Primary and secondary metabolites

The end products extracted from a fermentation process may be

- the entire micro-organism, such as mycoprotein for food products
- **primary metabolites** – substances produced by the microbe as part of its basic metabolism, such as ethanol and some enzymes
- **secondary metabolites** – substances the micro-organism makes only at a particular stage in its life cycle, or substances that the microbe has been artificially encouraged to produce from another compound. Antibiotics are secondary metabolites.

Aseptic techniques

Microbes such as bacteria can be grown in a **growth medium** containing **nutrients**, **buffers** to provide an appropriate pH and with **aeration** to provide **mixing** and, where necessary, **oxygen**.

The growth medium and the containers are **sterilised** to avoid infection by competitive or dangerous microbes.

The medium may be **liquid** or mixed with **agar** to form a **gel** that can be spread inside **petri dishes**.

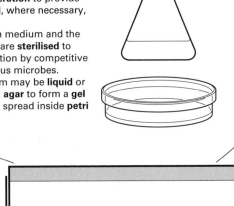

The dish is
- sealed with tape
- placed upside down (to prevent any condensation falling back onto the agar).
- in an incubating oven: 25°C allows growth of many microbes but not those which colonise the human body.

Any labelling should be on the bottom of the plate, just in case the lids become mixed up!

Aseptic techniques
- minimise danger to humans
- minimise contamination of cultures.

Microbes can be collected from any source, e.g.
- food, such as milk
- water samples
- skin
- pure cultures, obtained from microbiological suppliers, and transferred to a fresh, sterile agar plate.

Transfer of microbes: uses an **inoculating loop**, made of wire held in a steel handle.

Loop, and top of culture vessel, are held in Bunsen flame to sterilise them.

The transferred microbes are often spread onto a sterile agar plate in a **streak pattern**. The lid should be removed at the last moment to reduce the risk of contamination by airborne microbes.

Counting cells in a population of microbes

Total count: records all cells, whether they are alive (so capable of reproduction) or dead.

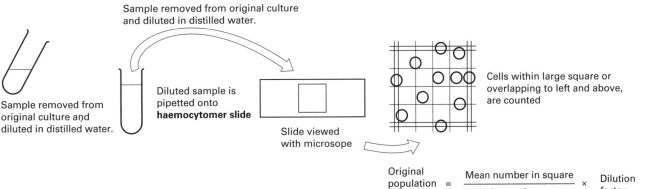

Sample removed from original culture and diluted in distilled water.

Sample removed from original culture and diluted in distilled water.

Diluted sample is pipetted onto **haemocytomer slide**

Slide viewed with microsope

Cells within large square or overlapping to left and above, are counted

$$\text{Original population density} = \frac{\text{Mean number in square}}{\text{Volume of square}} \times \text{Dilution factor}$$

Viable (live) count: records only cells that are **alive and capable of reproduction**.
A **serial dilution** is used to provide individual cells – these divide to produce colonies on agar plates.

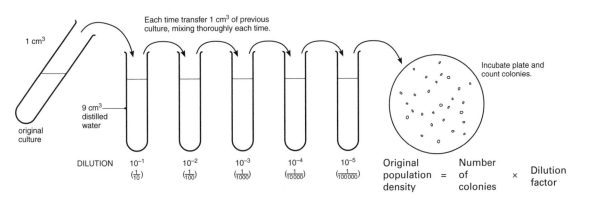

Each time transfer 1 cm³ of previous culture, mixing thoroughly each time.

1 cm³

9 cm³ distilled water

original culture

Incubate plate and count colonies.

DILUTION 10^{-1} $(\frac{1}{10})$ 10^{-2} $(\frac{1}{100})$ 10^{-3} $(\frac{1}{1000})$ 10^{-4} $(\frac{1}{10000})$ 10^{-5} $(\frac{1}{100000})$

$$\text{Original population density} = \text{Number of colonies} \times \text{Dilution factor}$$

Gene sequencing

Knowing the sequence of the human genome and advances in DNA technology enable us to identify and locate genes – particularly those that affect our risk of developing different diseases, and mutant genes which are the cause of genetic disorders.

The **chain-termination method** is one way of sequencing the bases of DNA. The diagram and its checklist are your guide to the technique.

Checklist

1 The DNA of entire chromosomes is fragmented using specific restriction enzymes.

2 The fragments are cloned using PCR.

3 Each fragment is sequenced.

4 The sequences are assembled into an overall sequence of the DNA of each chromosome.

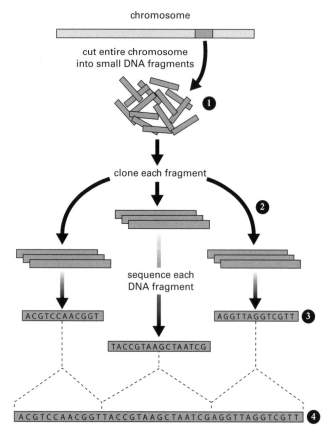

chromosome

cut entire chromosome into small DNA fragments

1

clone each fragment

2

sequence each DNA fragment

ACGTCCAACGGT   AGGTTAGGTCGTT **3**

TACCGTAAGCTAATCG

ACGTCCAACGGTTACCGTAAGCTAATCGAGGTTAGGTCGTT **4**

Whole genomes, including the human genome, have been sequenced using the chain termination method. The technology is highly automated and combined with sophisticated computer software. This makes it possible to assemble sequences of DNA fragments into a whole genome.

This knowledge of the genome of different individuals and of different species allows geneticists to compare genes and genomes between individuals and between species. A close genetic correlation suggests a close evolutionary relationship between species.

Discovering gene cloning technologies

1953 James Watson and Francis Crick propose a structure for DNA.

1961 Francois Jacob and Jacques Monod propose the operon model to explain the regulation of gene expression.

1966 Francis Crick and co-workers decipher genetic code.

1970 Enzymes are discovered which are a 'toolkit' for gene cloning technology.
- **Restriction endonucleases** (restriction enzymes) are 'molecular scissors' which catalyse reactions that cut strands of DNA into shorter lengths.
- **Ligases** catalyse the insertion (splicing) of lengths of DNA from one organism into the DNA of another different type of organism.
- **Reverse transcriptases** catalyse the synthesis of DNA from RNA.

1973 DNA is first inserted (spliced) into a **plasmid**. The technique allows genes to be cloned *in vivo*.

1977 Frederick Sanger describes the order of the bases (sequence) of the DNA of a virus called phiX174.

1983 Kary Mullis develops the **polymerase chain reaction** (**PCR**) which enables genes to be copied rapidly. The technique allows genes to be cloned *in vitro*.

1984 Alec Jeffreys and co-workers discover the technique of genetic **fingerprinting**.

Electrophoresis

The process of **electrophoresis** separates charged particles according to their molecular mass (size). For example, fragments of DNA carry a negative charge. The mixture of fragments to be separated is placed in wells at one end of a block of gel placed between two electrodes.

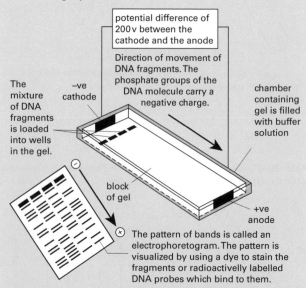

potential difference of 200 v between the cathode and the anode

Direction of movement of DNA fragments. The phosphate groups of the DNA molecule carry a negative charge.

The mixture of DNA fragments is loaded into wells in the gel.

−ve cathode

chamber containing gel is filled with buffer solution

block of gel

+ve anode

The pattern of bands is called an electrophoretogram. The pattern is visualized by using a dye to stain the fragments or radioactively labelled DNA probes which bind to them.

When the current is switched on separation of the DNA fragments begins. The fragments move through the gel towards the positive electrode. The smaller fragments move more quickly than the larger fragments and therefore travel greater distances. The different distances which the fragments move separate them into bands.

Enzymes: a toolkit for gene technology

Different types of enzyme enable us to obtain desirable genes (the genes we want), which we can manipulate to produce substances useful to us.

Restriction endonucleases

Each of the many different restriction endonucleases is named after the bacterium in which it occurs. For example *Eco* RI comes from the bacterium *Escherichia coli*; *Hind* III from *Haemophilus influenzae*. 'One' identifies Eco RI as the first restriction endonuclease isolated from *E.coli*; 'three' the third restriction endonuclease isolated from *H influenzae*.

The term 'restriction enzymes' will be used to refer to restriction endonucleases from now on. The enzymes catalyse the hydrolysis of the phosphodiester bonds linking the sugar/phosphate groups of each strand of a double-stranded molecule of DNA. Each one cuts at a point following or within a particular short sequence of base pairs. The short sequence is called the **recognition site**. The recognition site is **specific** to each restriction enzyme. For example the diagram represents the recognition site for *Eco* RI.

Notice that the recognition sequence is a palindrome... so called because the sequence GAATTC is the same as its reverse complement of bases CTTAAG.

As a result the fragments produced each have **sticky ends**.

The exposed bases of sticky ends are 'sticky' because they form hydrogen bonds with the complementary bases of the sticky ends of the fragments of other DNA molecules cut by the *same* restriction enzyme.

Not all restriction enzymes produce fragments of DNA each with sticky ends. For example, the restriction enzyme *Hpa* I cuts at the same place within its recognition site, producing **blunt-ended (flush-ended)** fragments.

The lengths of DNA produced as a result of the activity of restriction enzymes form a mixture of fragments called **restriction fragment length polymorphisms** – RFLPs or 'rif-lips' for short. The fragments can be separated according to size by **gel electrophoresis**. The fragment which contains the desirable gene may be identified using a **gene probe**. A gene probe (also called DNA probe) is a single strand of DNA containing a particular sequence of bases. The bases pair with the complementary bases in the gene under investigation. The probe is labelled with a fluorescent or radioactive label, so binding of the probe can be detected, revealing the presence of the gene. Once identified, cutting out the gene containing fragments from the **gel** makes them available for transfer into the DNA of another organism.

Ligase

A gene-carrying piece of DNA can be spliced (inserted) into another piece of DNA. The pieces bond with one another. After bonding, gaps called **nicks** are left. These nicks may be sealed by adding a phosphate group. The enzyme **ligase** catalyses the reaction.

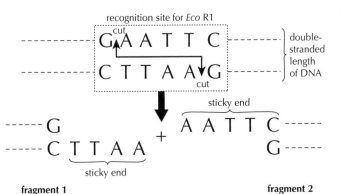

Recognition site for *Eco* RI

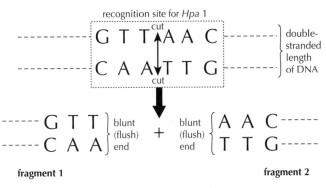

Flush-ended fragments

Reverse transcriptase

In 1957 Francis Crick suggested an idea he called the central dogma:

$$\text{DNA} \xrightarrow{\text{transcription}} \text{mRNA} \xrightarrow{\text{translation}} \text{polypeptide}$$

The arrows represent the transfer of information in the form of sequences of bases (DNA, mRNA) into the amino acid sequence of a polypeptide. However, the discovery that the genetic material of certain viruses is not DNA but RNA means that the central dogma does not apply in all cases. The viruses (called retroviruses, e.g. HIV) produce the enzyme **reverse transcriptase** which catalyses the conversion:

$$\text{RNA} \xrightarrow{\text{reverse transcriptase}} \text{cDNA}$$

... the reverse of the central dogma. The DNA produced is called **copy DNA (cDNA)** and does not carry sticky ends.

Questions

1 Briefly explain the role of restriction endonucleases and ligases.
2 Explain the process of gel electrophoresis.
3 The chain-termination method is one way of sequencing the bases of DNA. Summarize the method.
4 What is a recognition site?
5 What are sticky ends?

Plasmids and other vectors

Plasmids are circular pieces of DNA that occur in bacteria separate from the bacterial chromosome. Bacteria in a population can naturally exchange plasmids with each other, so exchanging the characteristics the DNA codes for. The advantage of this to the bacteria is that plasmids may carry genes that provide resistance to naturally occurring antibiotics, they may code for toxins, or they may provide bacteria with the ability to fix nitrogen or break down particular food molecules. They are replicated in the bacterial cell independently of the bacterial chromosome. Plasmids naturally enter host cells and are expressed there, so they make ideal vectors for genetic engineering.

Plasmids are not the only vectors used in genetic engineering:

- In experimental gene therapy for cystic fibrosis, **liposomes** in an aerosol spray carry the normal allele into the lungs, where cells take up the liposomes and express the normal allele.
- When a virus infects a host, **viral DNA** is naturally inserted into the host cell's DNA, so viral DNA can be engineered to carry to new gene into the host.
- **Cosmids** are recombinant vectors that combine features of both plasmids and virus DNA. They reproduce in the cell separately from the chromosomes in the same way as plasmids.

In **genetic engineering**, manufactured genes or genes extracted from the cells of one type of organism are transferred to the cells of almost any other type of organism: plants → bacteria; humans → bacteria; bacteria → plants; and so on. The cells of the organism into which genes are transferred are the **host** cells and are said to be **transformed**. The organism itself is **genetically modified (GM)** and it expresses the new gene – it produces the protein the gene codes for.

Vectored transfer

A **vector** is a piece of DNA into which a desirable gene can be inserted. The result is a mixed (**hybrid**) molecule consisting of vector DNA and the desirable gene. The term **recombinant DNA** refers to the hybrid molecule which itself is said to be **genetically engineered**.

Inserting the desirable gene into a DNA vector is possible because of sticky ends. If the fragment of DNA containing the desirable gene is cut from the host DNA by a particular restriction enzyme which produces sticky ends, and the same restriction enzyme is used to cut the vector DNA, then the sticky ends of the gene-carrying fragment of DNA and the vector DNA are complementary and bond with one another. Ligase is used to seal the nicks (gaps) left after bonding.

Sticky ends can be added to strands of cDNA and DNA cut into blunt-ended fragments by restriction enzymes which do not produce sticky ends. Producing the sticky ends is possible by mixing the blunt-ended fragments of DNA with a mixture of nucleotides carrying the bases A T C or G and **terminal transferase**. The enzyme catalyses the addition of short single-stranded sequences of the nucleotides to each end of a blunt-ended fragment.

Bacterial plasmids are commonly used vectors. For example, a plasmid of the soil bacterium *Agrobacterium tumefasciens* induces cell division in plant tissues infected with the bacterium. Swellings called **galls** develop. Each swelling is a type of tumour. The condition is known as **crown gall disease** and the plasmid is called the **Ti (tumour inducing)** plasmid. If a desirable gene is inserted into the Ti plasmid, and a plant infected with bacteria carrying the recombinant material, the cells in the gall which develops each contain the Ti plasmid with the desirable gene in place. Multiplication of the cells in small pieces of tissue (explants) cut out of the gall produce many copies of the Ti plasmid and the desirable gene it carries – an example of **gene cloning** *in vivo*. The term *in vivo* literally means 'within the living'. We say that genes are cloned *in vivo* when they are replicated in cells dividing by mitosis.

Plantlets develop from explants placed in growth medium to which plant growth regulators have been added. Once transferred to soil, the plantlets grow into plants. The plants are **genetically modified (GM)** because they contain the recombinant Ti plasmid. Crops containing a herbicide-resistant gene were first produced in this way.

Engineering *Agrobacterium tumefasciens*

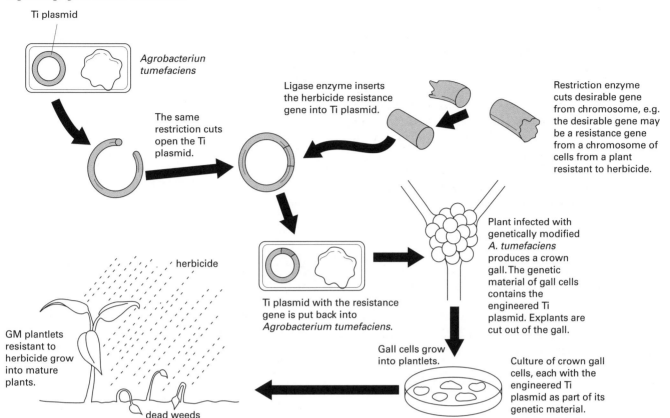

Ti plasmid

Agrobacteriun tumefaciens

Ligase enzyme inserts the herbicide resistance gene into Ti plasmid.

Restriction enzyme cuts desirable gene from chromosome, e.g. the desirable gene may be a resistance gene from a chromosome of cells from a plant resistant to herbicide.

The same restriction cuts open the Ti plasmid.

Plant infected with genetically modified *A. tumefaciens* produces a crown gall. The genetic material of gall cells contains the engineered Ti plasmid. Explants are cut out of the gall.

herbicide

Ti plasmid with the resistance gene is put back into *Agrobacterium tumefaciens*.

Gall cells grow into plantlets.

Culture of crown gall cells, each with the engineered Ti plasmid as part of its genetic material.

GM plantlets resistant to herbicide grow into mature plants.

dead weeds

Vectorless transfer

Protoplasts are plant cells stripped of their cell walls. The diagram shows how protoplasts are prepared.

- Protoplasts will take up DNA added to the medium in which they are growing. They divide and treatment with plant growth substances stimulates the development of whole plants which carry the genes taken up by the original protoplasts.

- *Gene guns* deliver DNA directly into plant tissue. Tiny pellets of gold or tungsten coated with DNA are fired at high velocity through the walls of the cells of the tissues placed in their path. Some of the cells take up the DNA. The technique is called **biolistics**.

plant tissue cut from the whole plant

the tissue is treated with sodium chlorate(I) (or other suitable sterilant) solution to kill bacteria.

The tissue is treated with **cellulase**. The enzyme catalyses the hydrolysis of cellulose in the cell walls which are removed.

suspension of protoplasts

- *Electroporation* involves delivering bursts of electricity to cells growing in liquid containing donor DNA. The electric field creates temporary pores in the plasma membrane surrounding each cell, allowing the donor DNA to enter the cells.

- *Microinjection* involves using a syringe with a very fine glass-needle to inject DNA into the nucleus of a host cell.

In each case the transferred genes are cloned *in vivo* when the host cells divide.

Questions

1 Why is recombinant DNA said to be a hybrid molecule?

2 Summarize the arguments for and against GM crops.

3 Explain how the vectorless transfer of genes into host cells is possible.

4 Why might cloning mammals be controversial?

Forming multiple copies of DNA fragments is possible using the **polymerase chain reaction (PCR)**. The DNA is cloned: we say that it is **amplified**.

Any fragment of DNA, including genes, may be cloned using PCR. Millions of copies of a DNA fragment can be synthesized in a few hours. The cloning process does not involve living cells. It is performed in labware. We refer to the technique as *in vitro* cloning.

The term *in vitro* literally means 'within the glass'. However, cloning genes *in vitro*, for example, is not necessarily performed in test tubes. The process may be contained in Petri dishes.

PCR is widely used in biological and medical research. For example, it can be used to determine evolutionary relationships between organisms, and speed up the diagnosis of infectious diseases, genetic disorders, and different types of cancer.

Often only traces of DNA evidence are found at the scene of a crime. However, amplifying the DNA collected using PCR makes enough material available for analysis.

Relative advantages of *in vivo* and *in vitro* cloning

Testing drugs on animals and clinical trials which test the safety of new drugs on human volunteers are examples of *in vivo* techniques. Variables are more easily controlled *in vitro*. The relative advantages of *in vivo* and *in vitro* techniques are summarized in the table.

Overall, *in vitro* techniques are often simpler, cheaper, and more sensitive than techniques in *vivo*. However conditions *in vitro* do not correspond to real life conditions *in vivo*, and may produce misleading results. *In vitro* studies therefore, are usually followed up *in vivo*.

In vivo techniques	*In vitro* techniques
Disadvantages • DNA cloning may take several weeks • a relatively large amount of DNA is needed for cloning • high cost • only fresh DNA can be cloned successfully	**Advantages** • DNA cloning is complete in a few hours • only the minutest amount of DNA is needed for cloning • low cost • degraded DNA from fossils can also be cloned
Advantages • long sequences of DNA up to 2 Mb in size can be cloned • providing the conditions promoting cell division are maintained, the amount of DNA cloned is potentially limitless	**Disadvantages** • only short sequences of DNA can be cloned • in theory the amount of DNA cloned is limitless; however in practice less DNA is cloned

The polymerase chain reaction (PCR). The cycle is repeated (25 cycles generates >1 million copies of the original double-stranded DNA). Each cycle lasts about 2 minutes.

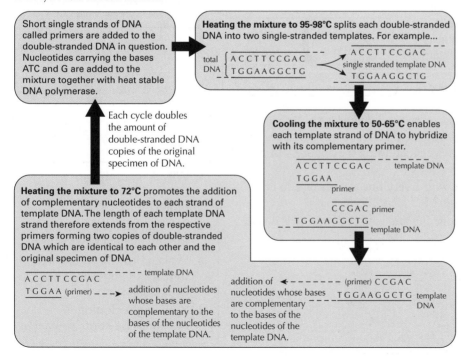

The size of DNA sequences

Gb = giga base pairs (1 thousand million)

Mb = mega base pairs (1 million)

kb = kilo base pairs (1 thousand)

The human genome is 3.1 Gb.

Questions

1 What is the difference between the *in vivo* and *in vitro* cloning of genes?

2 List some of the applications of PCR.

3 A genome consists of 2700 million base pairs. Which of the following units correctly describes its size?

a 2.7 kb   **b** 2.7 Gb   **c** 2.7 Mb

Making genetically engineered insulin

People with type I diabetes inject insulin to help regulate the concentration of glucose in their blood.

The human insulin gene can be synthesized or cut from the chromosome carrying it and inserted into a plasmid vector. The vector carries the gene into the cells of the bacterium *Escherichia coli*. The modified bacterium makes insulin. The diagram summarizes the process.

Notice:

❶ The process starts either with the mRNA encoding human insulin or the original gene. The ß cells of the Islets of Langerhans are rich in the mRNA encoding insulin, so its extraction is straight forward. The extracted mRNA is used to synthesize cDNA. The synthesis is catalysed by the enzyme reverse transcriptase.

❷ The plasmid vector carries genes encoding resistance to different antibiotics. The resistance genes are important as genetic markers when it comes to identifying colonies of transformed bacteria carrying the human insulin gene.

❸ Once the plasmid vector carrying the insulin gene is in place in *E.coli* cells, we say that the bacteria are genetically modified (GM). The gene continues to encode the synthesis of insulin because the genetic code works the same way in all cells.

❹ The genetically modified bacteria are put into a solution containing all of the nutrients they need for rapid replication. Replication of the bacterial cells replicates the plasmid vector and the gene it carries. The gene is cloned *in vivo*.

❺ The solution of nutrients fills huge containers called **fermenters**. The insulin produced by the bacteria is separated from the solution, purified and then packaged ready for sale (downstream processing).

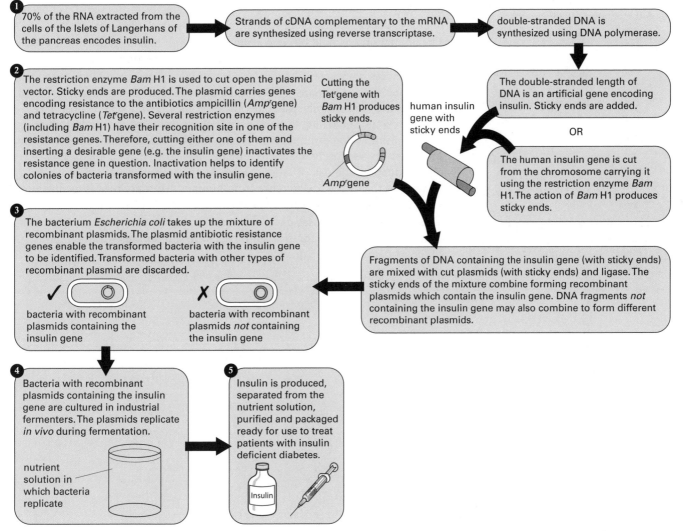

❶ 70% of the RNA extracted from the cells of the Islets of Langerhans of the pancreas encodes insulin.

Strands of cDNA complementary to the mRNA are synthesized using reverse transcriptase.

double-stranded DNA is synthesized using DNA polymerase.

❷ The restriction enzyme *Bam* H1 is used to cut open the plasmid vector. Sticky ends are produced. The plasmid carries genes encoding resistance to the antibiotics ampicillin (*Amp'*gene) and tetracycline (*Tet'*gene). Several restriction enzymes (including *Bam* H1) have their recognition site in one of the resistance genes. Therefore, cutting either one of them and inserting a desirable gene (e.g. the insulin gene) inactivates the resistance gene in question. Inactivation helps to identify colonies of bacteria transformed with the insulin gene.

Cutting the *Tet'*gene with *Bam* H1 produces sticky ends.

*Amp'*gene

human insulin gene with sticky ends

The double-stranded length of DNA is an artificial gene encoding insulin. Sticky ends are added.

OR

The human insulin gene is cut from the chromosome carrying it using the restriction enzyme *Bam* H1. The action of *Bam* H1 produces sticky ends.

❸ The bacterium *Escherichia coli* takes up the mixture of recombinant plasmids. The plasmid antibiotic resistance genes enable the transformed bacteria with the insulin gene to be identified. Transformed bacteria with other types of recombinant plasmid are discarded.

✓ bacteria with recombinant plasmids containing the insulin gene

✗ bacteria with recombinant plasmids *not* containing the insulin gene

Fragments of DNA containing the insulin gene (with sticky ends) are mixed with cut plasmids (with sticky ends) and ligase. The sticky ends of the mixture combine forming recombinant plasmids which contain the insulin gene. DNA fragments *not* containing the insulin gene may also combine to form different recombinant plasmids.

❹ Bacteria with recombinant plasmids containing the insulin gene are cultured in industrial fermenters. The plasmids replicate *in vivo* during fermentation.

nutrient solution in which bacteria replicate

❺ Insulin is produced, separated from the nutrient solution, purified and packaged ready for use to treat patients with insulin deficient diabetes.

Insulin

Why are GM crops an issue?

GM crops would seem to have real benefits. For example, growing them helps farmers to control the weeds and insect pests that damage crops and therefore reduce food production. Growing them also cuts back on the chemicals used to control the numbers of weeds and insect pests. This results is less damage to wildlife and the environment.

Why, then, are people concerned about the growing of GM crops? Here are some points to think about:

- People are worried that eating GM food may harm their health.

- There are concerns that GM crops harm wildlife.

- Pollen from crops genetically modified to resist herbicides may transfer to wild plants. If these plants are weeds, there is a danger of the development of weeds resistant to herbicides.

- Some people think that transferring genes between organisms in the laboratory is somehow 'not natural'.

- Developing countries are less easily able to afford the science needed to develop GM crops. However many of them are countries most likely to benefit from the technology. Biotech companies in developed countries can exploit the demand and dominate global markets in places least able to buy in the benefits. Anti-globalization activists believe it wrong that developed countries can dominate markets because of their financial muscle.

Remember that whatever your views, there are different ways of thinking about the issues raised.

Golden Rice™

Golden Rice™ is a product of research in which rice has been genetically engineered to contain beta-carotene, a precursor of vitamin A, in the endosperm, which is a tissue in the rice grains. Beta-carotene is not present in the endosperm of ordinary white rice grains. The human body naturally converts beta-carotene into vitamin A. Rice is the staple diet in many parts of the world and eating Golden Rice prevents deficiency diseases from lack of vitamin A in the diet, including blindness in children which is widespread in some areas. The research aims to produce enough beta-carotene in Golden Rice to cover the recommended daily requirements.

The rice plant can naturally produce beta-carotene in its leaves. Golden Rice was created by genetic engineering in which beta-carotene biosynthesis genes were added. A promoter was added so that these genes are only expressed in the endosperm. The end product of the engineered pathway is a substance that would turn the rice red, but the plant's natural enzymes process this substance in the endosperm, giving the rice a distinctive yellow colour. The resulting rice plant was cross-bred with local varieties in the Philippines and Taiwan.

It is hoped that the rice will pass stringent regulatory requirements and be widely available to developing countries as part of a humanitarian project.

Xenotransplantation

Xenotransplantation is the transplantation of living cells, tissues, or organs from one species to another. Xenotransplantation into human recipients may provide treatment for organ failure such as kidney or heart failure, a significant health problem in parts of the industrialized world. A few xenotransplantation procedures have been successful.

There is a shortage of human organs for transplant, and many patients die while waiting for a suitable organ. Procedures are being investigated which use cells or tissues from other species to treat serious illnesses such as cancer, diabetes, liver failure, and Parkinson's disease. Long-term storage of xenogenic organs is under investigation, so that they can be available for transplant.

The animal organ could be genetically altered with human genes to trick the patient's immune system into accepting it without rejection.

Chimpanzees have organs of similar size, and they have good blood type compatibility with humans. However, chimpanzees are an endangered species. Baboons are more readily available; however, they have a smaller body size, poor blood type matching to humans, a long gestation period, and produce few offspring. Using primates carries an increased risk of disease transmission, since they are so closely related to humans. Pigs are better candidates for organ donation: the risk of cross-species disease transmission is decreased because of their genetic difference from humans. They are readily available and their organs are comparable in size.

Xenotransplantation raises many medical, legal, and ethical issues. Apart from concerns about disease transmission, pigs have shorter lifespans than humans, so their tissues age more quickly.

Gene therapy

We all carry a few mutant alleles (*recall* that alleles are pairs of genes encoding a particular polypeptide). Most of them are recessive, so their possible harmful effects are masked by their normal dominant partners. However, when recessive mutations are homozygous or when the mutation is dominant, then individuals carrying them are at risk of genetic disorders. The particular disorder depends on the particular mutation.

Gene therapy aims to supplement mutant alleles with normal ones. However, the techniques are difficult, although research into long-term cures for genetic disorders is making progress.

Remember, for example, that cystic fibrosis is a genetic disorder caused by a mutant recessive allele. Healthy copies of the allele are engineered into liposomes. These are carried as an aerosol spray deep into the lungs of a person affected by cystic fibrosis in an attempt to supplement the activity of the defective gene. The cells seem to take up the liposomes each with the healthy allele. However, the healthy allele must become part of the person's DNA at a locus where it is effective. Finding the right locus is just one of the difficulties of gene therapy. So far the relief of symptoms of cystic fibrosis is usually temporary. Research continues to try to make the benefits long term.

This experimental treatment for cystic fibrosis is an example of **somatic line therapy**. It aims to alter the body (somatic) cells so that they produce the normal protein. **Germ line therapy** in contrast aims to alter the gametes – the egg or sperm is engineered before fertilisation so that the altered gene is passed on to future generations. Human germ line therapy is not allowed in the UK.

Recombinant DNA and animal cloning: some issues

Some people think the benefits of recombinant DNA technology are potentially enormous; others have serious moral and ethical concerns. Here are a few points for you to think about.

Recombinant DNA technology is well established. Cloning animals from cells grown in culture is also possible. Dolly the sheep is an example. If we can clone animals like Dolly, then cloning human beings should be relatively simple (although currently illegal worldwide).

The cells which gave rise to Dolly were not transformed: they did not contain recombinant DNA. However, the technology to clone animals from transformed cells is available. Many people think that the theoretical possibility of cloning humans from transformed cells is a very dangerous idea. Others argue that such a development would make it possible to eliminate many genetic disorders.

Remember that whatever your views there are different ways of thinking about the issues raised.

Questions

1 Why might it be better to synthesize the human insulin gene for insertion in *E. coli* rather than isolate it from human genetic material?

2 Why does the bacterial plasmid engineered with the human insulin gene also carry genes encoding resistance to different antibiotics?

2.22 Populations and ecosystems

Key terms

- Biosphere
- Community
- Ecosystem
- Habitat
- Niche
- Population

All of the places on Earth where there is life form the biosphere. Organisms are adapted (suited) to where they live. Where they live is made up of:

- an **abiotic** (non-living) environment of air/soil/water
- a **biotic** (living) **community** of animals, plants, fungi, and microorganisms.

An environment and its community form an **ecosystem** which is a more or less self-contained part of the biosphere. 'Self-contained' means that the organisms living in a particular environment are characteristic of that environment because of the adaptations that enable them to live there and not in a different ecosystem. Put simply, for example, roots anchor plants in soil; fins enable fish to swim in water. Ecosystems are dynamic – they are constantly changing as their biotic and abiotic factors change.

Each type of organism (species) in an ecosystem forms a **population**. The term **ecology** refers to the relationship between populations and between populations and the environment of the ecosystem of which they are a part. The other terms listed in the box (see left) refer to the different components of all the ecosystems which make up the biosphere.

The diagram shows the relationship between the terms in an oakwood ecosystem. The numbers on the diagram refer to the components of the ecosystem and are summarized in the checklist points. The organization of the components is the same for any ecosystem.

COMMUNITY ❷

Key 1 oak tree
2 hazel
3 holly
4 bluebell
5 wood anemone
6 primrose
7 moss on tree trunk
8 pigeons, rooks living in canopy
9 bluetits, woodpeckers living further down tree
10 great tits, warblers living in shrubs
11 wrens, blackbirds living on ground
12 toadstools on rotting log
13 woodlice in detritus
14 earthworm pulling leaf into burrow falling leaves

Fact file

What are adaptations?

Adaptation refers to all of the characteristics of an organism which enable it to survive in a particular environment. Characteristics include an organism's:

- **morphology** – body structure
- **molecular biology** – the organization and function of molecules in cells
- **physiology** – the way the body works
- **biochemistry** – the chemistry of cells
- **behaviour** – an individual's reactions to changes in its environment

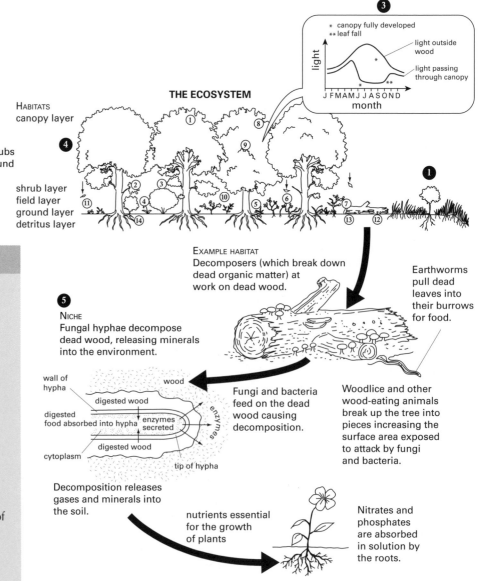

HABITATS
canopy layer

shrub layer
field layer
ground layer
detritus layer

THE ECOSYSTEM

* canopy fully developed
** leaf fall

light outside wood

light passing through canopy

light

J F M A M J J A S O N D
month

EXAMPLE HABITAT
Decomposers (which break down dead organic matter) at work on dead wood.

Earthworms pull dead leaves into their burrows for food.

❺
NICHE
Fungal hyphae decompose dead wood, releasing minerals into the environment.

wall of hypha
digested wood
digested food absorbed into hypha
enzymes secreted
digested wood
cytoplasm
tip of hypha
wood
enzymes

Fungi and bacteria feed on the dead wood causing decomposition.

Woodlice and other wood-eating animals break up the tree into pieces increasing the surface area exposed to attack by fungi and bacteria.

Decomposition releases gases and minerals into the soil.

nutrients essential for the growth of plants

Nitrates and phosphates are absorbed in solution by the roots.

The components of an oak wood ecosystem

94

Checklist

1 • There is overlap where the boundaries of ecosystems meet. For example, the boundary of an oakwood is where the trees thin into grassland.
  - As a result organisms from each ecosystem may be found at the boundary.
  • The exchange of organisms across the boundary is limited because the organisms are **adapted** to the particular ecosystem in which they live.
  - As a result the unique character of each ecosystem is maintained.

2 • Individuals of a particular type of organism (**species**) living in a particular place at a particular time, form a **population**. The **community** is made up of all of the different populations living in a particular ecosystem at a particular time.

3 • The **environment** is the place where a community of organisms lives.
  • Different factors in the environment affect the distribution and types of organism of a community:
    ◦ **abiotic** (physical) factors include climate, conditions of air/soil/water, characteristics of location such as altitude (the height of land above sea level) and depth of water
    ◦ **biotic** factors include competition, predator/prey interactions, other relationships between organisms

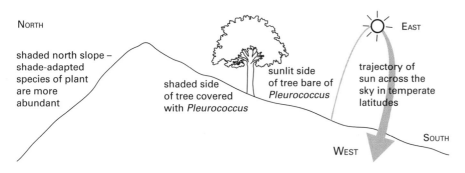

NORTH

shaded north slope – shade-adapted species of plant are more abundant

shaded side of tree covered with *Pleurococcus*

sunlit side of tree bare of *Pleurococcus*

EAST

trajectory of sun across the sky in temperate latitudes

WEST

SOUTH

Aspect (lie of the land) is an **abiotic** factor which affects the distribution of organisms. *Pleurococcus* is a type of unicellular green alga which grows on the bark of trees. It is more abundant on the shaded north side of tree trunks than on the more brightly lit south side.

4 • The **habitat** is the localized part of the environment where a population lives and which provides most of the resources its members need.
  • *Notice* that the different layers of vegetation in the oakwood provide habitats for different bird species, for example.

5 • The **niche** is the totality of all that an organism does in its habitat, including all of the resources it consumes. Resources refer to food, space, availability of mates, and other requirements which enable a species to survive in its habitat. Information about its niche tells us what a species feeds on, what feeds on it, where it rests, and how it reproduces.
    ◦ **Fundamental niche** – all of the resources a species can exploit in theory in the absence of competition.
    ◦ **Realized niche** – all of the resources a species can exploit in reality because of competition.
  • *Notice* on the concept map the example habitat of a rotting log. Fungi and bacteria feed on the dead wood causing its decomposition. They are called **decomposers**. Their niche in the habitat breaks down dead organic matter.
    - As a result the decomposers obtain food and their activities release different compounds as gases and nutrients into the environment.
    - As a result elements (e.g. carbon and nitrogen) are recycled through the ecosystem.

Fact file

In 1934 the Russian biologist G F Gause stated that a niche is occupied by only **one species population** at any one time. The principle is called the **Gause hypothesis**.

- As a result competition for limited resources between different species is reduced.

- As a result niches are **separated** from each other and limited resources are shared between different species.

- As a result resources in limited supply are not over-exploited.

Fact file

The cells of most types of fungi consist of a mass of thread-like cells, forming a **mycelium**. Each thread is called a **hypha** (pl. hyphae).

Questions

1 Explain your understanding of ecology.
2 What is the difference between a habitat and niche?
3 Explain your understanding of adaptations.

2.23 Food chains, food webs, and ecological pyramids

The community works through the interactions between its different niches. *Remember* that **food** is an important part of the niche idea.

⊙ As a result describing feeding relationships is one way of finding out how a community works.

Food chains and food webs

Different words are used to describe the feeding relationships between the plants and animals of a community.

- **Producers** make food (sugars). Plants are producers. So too are some types of single celled organisms and algae. They make food by **photosynthesis**. Food chains and food webs always begin with producers.
- **Consumers** take in food (feeding) already formed. Animals are consumers.

A **food chain** shows the links between producers and consumers. It describes one pathway of food through a community. For example:

oak leaves ⟶ slugs ⟶ thrush ⟶ sparrowhawk

A **food web** describes all possible feeding relationships of a community. It consists of interlinked food chains. The diagram shows the food web in an oakwood.

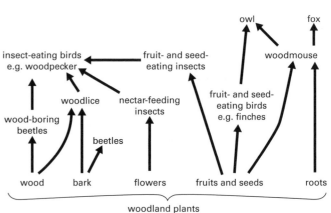

Ecological pyramids

Food webs represent the feeding relationships of communities but do not indicate the numbers of individuals involved. The description is **qualitative**. Ecological pyramids represent a **quantitative** description of feeding relationships. They describe *how much* food is transferred through the community. *Notice* that organisms with *similar* types of food are grouped into **trophic** (feeding) **levels**.

- The producer trophic level occupies the base of the pyramid.
- Other trophic levels are made up of consumers:
  - **primary** consumers are herbivores (H)
  - **secondary** consumers or first/primary carnivores (C_1) feed on herbivores
  - **tertiary** consumers or second/secondary carnivores (C_2) feed on first/primary carnivores

There are different types of ecological pyramid.

- **Pyramids of numbers** represent the number of organisms in each trophic level at a particular time. The diagram compares the pyramid of numbers for a grassland community and a woodland community.

Notice that for the grassland community

- a lot of producers (grass plants) support many herbivores (e.g. insects), which in turn support fewer carnivores (e.g. spiders).

⊙ As a result the pyramid is upright and tapers to a point.

Notice that for the woodland community

- few producers (trees) support many herbivores (e.g. insect larvae), which in turn support fewer carnivores (e.g. birds, spiders).

⊙ As a result the pyramid has a narrow base.

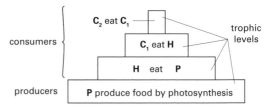

Key
P = plants H = herbivores C_1 = first carnivores
C_2 = second carnivores

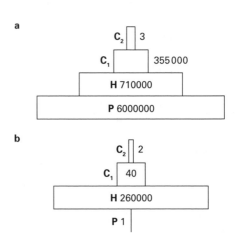

Pyramids of numbers for **a** grassland and **b** woodland communities

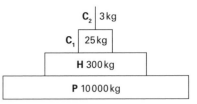

What is the problem?

The producers and consumers in the grassland community are mostly small, so differences in size are slight and do not have a marked effect on the shape of the pyramid. However the pyramid of numbers for the woodland community does not take into account the big differences in size of the producers and consumers. Each tree is large and can meet the food needs of many smaller consumers.

- **Pyramids of biomass** represent the amount of organic material in each trophic level (**standing crop** or **standing crop biomass**) at a *particular time*.

The diagram shows the pyramid of biomass for a woodland.

Notice that the pyramid is upright and tapers to a point. Measuring the standing crop in each trophic level allows for differences in size of producers and consumers.

But there are still problems!

- The biomass of an organism can vary during the year. For example, a tree in full leaf during the summer has a much greater biomass than in winter without its leaves.
- The diagram shows a pyramid of biomass for the English Channel. *Notice* that the pyramid points down.
    - The producers do not live as long as the herbivores which depend on them but reproduce quickly.
    - The rapid turnover of producers means that, for a time, the standing crop of producers may be smaller than the herbivores.
    - No account is taken of the *rate* at which biomass is produced or consumed. Only the amount of organic material in each trophic level at a *particular time* (standing crop – see above) counts in a pyramid of biomass.

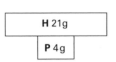

- **Pyramids of energy** take account of the amount of organic material produced and consumed over a *period of time*. The organic matter represents a store of energy. This energy can be estimated by burning the material in a bomb calorimeter and measuring the temperature rise. The diagram shows a pyramid of energy for a stream. It illustrates the energy flow through, and energy loss, from the trophic levels of the community.

Notice:

- The energy that flows through trophic levels of the community begins with sunlight, which enables producers to produce food by photosynthesis.
- The food produced flows through the community because the consumers of a particular trophic level feed on the producers/consumers of the preceding trophic level.
    - As a result food energy is transferred from one trophic level to the next.
- At each trophic level some of the food energy is used by organisms to fuel their own metabolism. The energy is eventually lost as heat from the community.
    - As a result the transfer of energy between trophic levels is never 100% efficient.
    - As a result the amount of energy decreases as it flows from one trophic level to the next through the community.
    - As a result the pyramid of energy tapers to a point.

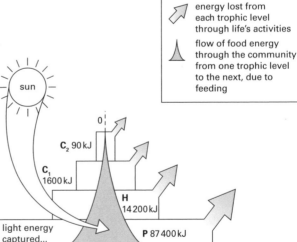

Pyramid of energy for a stream in kJ m^{-2} year^{-1}

But there are still problems with ecological pyramids!

- Many organisms feed at several trophic levels.
- Not all parts of a plant produce food or are available to herbivores. Roots are buried and do not contain chlorophyll. Yet the whole plant contributes to the data for the producer trophic level.
- The flow of food energy through decomposers is often omitted from pyramid diagrams.

Questions

1. What is a trophic level?
2. List the different types of consumer and briefly explain what each type eats.
3. Why is the pyramid of biomass usually a better description of feeding relationships in a community than a pyramid of numbers?

Light energy is trapped and converted by producers into the chemical bond energy of sugars by the reactions of photosynthesis. Energy flows through food chains and food webs when

- **herbivores** feed on plants
- **carnivores** feed on herbivores and other carnivores
- **decomposers** feed on dead organic material

In other words, feeding transfers energy between trophic levels: producers → consumers

Energy flow

Sunlight underpins the existence of most of life on Earth. Without sunlight and the producers which convert its energy into food energy by photosynthesis, most communities would not exist. The diagram tracks the flow of energy between trophic levels: producer → herbivore → carnivore. Follow the sequence of the checklist ❶ – ❺ which refer to the diagram. Together, the diagram and checklist will help you to quantify the efficiency of energy transfer between trophic levels.

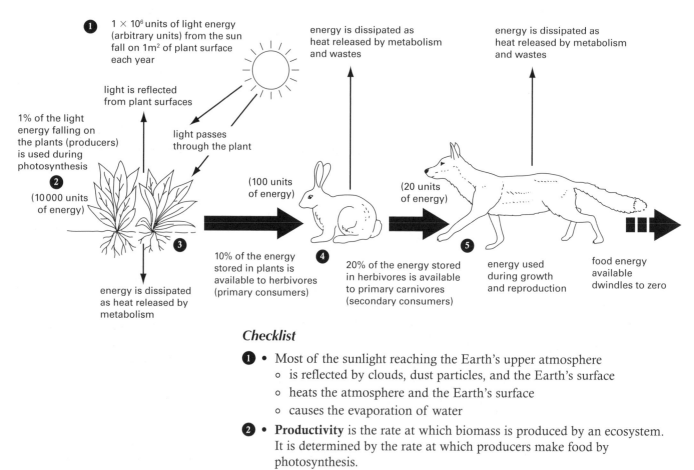

❶ 1×10^6 units of light energy (arbitrary units) from the sun fall on 1m² of plant surface each year

light is reflected from plant surfaces

1% of the light energy falling on the plants (producers) is used during photosynthesis

light passes through the plant

❷ (10 000 units of energy)

❸ energy is dissipated as heat released by metabolism

❹ 10% of the energy stored in plants is available to herbivores (primary consumers)

(100 units of energy)

energy is dissipated as heat released by metabolism and wastes

20% of the energy stored in herbivores is available to primary carnivores (secondary consumers)

❺ (20 units of energy)

energy used during growth and reproduction

energy is dissipated as heat released by metabolism and wastes

food energy available dwindles to zero

Checklist

❶ • Most of the sunlight reaching the Earth's upper atmosphere
  ○ is reflected by clouds, dust particles, and the Earth's surface
  ○ heats the atmosphere and the Earth's surface
  ○ causes the evaporation of water

❷ • **Productivity** is the rate at which biomass is produced by an ecosystem. It is determined by the rate at which producers make food by photosynthesis.
  • **Gross primary production (GPP)** refers to the biomass of food produced in g m⁻² (of plant surface exposed to sunlight) year⁻¹.
  • **Net primary production (NPP)** refers to GPP less the biomass used by consumers to fuel their own metabolism. NPP is used to compare the productivity of different ecosystems.

Fact file

Fragments of dead material (e.g. twigs, leaves) form **detritus**. Animals which feed on detritus are called detritivores.

Ecosystem	NPP g m^{-2} yr^{-1}
coral reefs	2600
tropical rain forests	2150
temperate deciduous forests	1300
temperate grasslands	500
open ocean	130
hot desert	65

3 • Some parts of plants may not be palatable (e.g. bitter tasting leaves) or accessible (e.g. roots) to herbivores.

     🅔 As a result the parts are not eaten and therefore not part of the food energy available to herbivores.

4 • Animals do not produce enzymes that digest cellulose and lignin.

     🅔 As a result plant material is difficult to digest.

• Herbivores depend on microorganisms living in the gut to produce enzymes which digest cellulose and lignin. The process is inefficient.

     🅔 As a result herbivores produce a large amount of faeces which contains undigested plant material, representing energy not available to carnivores.

5 • Animal material is easier to digest than plant material.

• Animal material has a higher energy value than plant material.

     🅔 As a result the transfer of energy from herbivore → carnivore is more efficient (20%) than producer → herbivore (5–10%).

• The indigestible parts of prey (hooves, hides, bones) represent energy not available to other carnivores.

Notice that energy is dissipated (dispersed) from the food chain to the environment at each stage of energy transfer between the sun and trophic levels: sun → producer; producer → primary consumer (herbivore); primary consumer (herbivore) → secondary consumer (primary carnivore)... and so on.

In summary:

• Assume that 1×10^6 units of solar light energy (arbitrary units) are available to the trophic levels of the food chain. *Notice* that only 20 units of chemical energy (as food) are available to primary carnivores.

     🅔 As a result the number of trophic levels is limited: usually 3 to 4.

• The more productive the ecosystem (the greater is the producer biomass), the more trophic levels there are: usually a maximum of 6.

Energy flow through decomposers

The flow of energy through decomposers and detritivores is often left out of diagrams representing food chains, food webs, and ecological pyramids. *Remember*, however, that in some communities, 80% of the productivity of a trophic level may flow through decomposer food chains.

• The energy locked up in the remains of dead organisms and wastes (faeces and urine) enters decomposer food chains where the activities of fungi and bacteria break down the organic material.

   ○ In tropical rain forests, the warm moist environment promotes the activities of decomposers.

     🅔 As a result the rate of decomposition is rapid.

     🅔 As a result little organic material accumulates in the ecosystem.

     🅔 As a result the soil of rain forests is nutrient poor.

   ○ In peat bogs the cold, wet, acidic environment inhibits the activities of decomposers.

     🅔 As a result the rate of decomposition is slow.

     🅔 As a result much organic material accumulates, forming **peat**.

Qs and As

Q Why are there a limited number of trophic levels in an ecological pyramid and links in a food chain?

A *The number of trophic levels and links in a food chain is limited by the food energy available. As the amount of food energy passing from one trophic level to the next decreases, so does the amount of living material that can be supported in each trophic level. When food energy dwindles to zero, trophic levels and links in food chains can no longer exist.*

Q Why is a pyramid of energy the best way of representing feeding relationships between organisms in different trophic levels?

A *Pyramids of numbers and biomass represent feeding relationships at a particular time. A pyramid of energy represents feeding relationships over a period of time. In other words, it takes account of the rate of production and consumption of biomass and deals with problems of*

• *differences in the size of organisms*

• *seasonal variations of biomass*

• *the rapid turnover of small organisms relative to the consumers that depend on them*

Questions

1 How does the description of the flow of energy through the community help us understand why there is a limited number of links in a food chain?

2 Suggest why the productivity of tropical rain forests is greater than temperate grasslands.

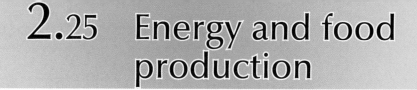

2.25 Energy and food production

Farms are ecosystems with people as consumers in a food chain of crops and livestock. Farmers manipulate the flow of energy through the ecosystem to maximise food production. The amount of food produced depends on

- energy *input*: sunlight and fuel oil
- energy *output*: efficiency of the conversion of energy input into the energy content of the food produced (crops and livestock)

The different practices which help to maximize the amount of food produced are what is called **intensive farming**.

Energy inputs and outputs have values. The diagram shows the energy values of different inputs and outputs for a modern intensive farm of 460 hectares in Southern England.

Photosynthetic efficiency is a measure of how well the plants of an ecosystem convert light energy into the chemical energy stored in the bonds of sugar molecules by photosynthesis.

$$\text{photosynthetic efficiency} = \frac{\text{energy content of plants year}^{-1}}{\text{light energy available year}^{-1}}$$

The greater the photosynthetic efficiency, the greater is the productivity (measured as net primary productivity) of the ecosystem.

- In natural ecosystems photosynthetic efficiency (and therefore productivity) varies depending on temperature, rainfall, and concentration of carbon dioxide.

  - As a result values for photosynthetic efficiency vary between 0.5% and 1% of the total energy value of sunlight reaching ground level.
- The photosynthetic efficiency of farm crops grown intensively may range up to 6%.

If we assume that the energy input of sunlight into farm ecosystems is the same as natural ecosystems…

then… the difference in photosynthetic efficiency (and productivity) between farm crops and the plants of natural ecosystems is due to the non-sun energy inputs into intensive farms. The inputs are based on fuel oil. More than 40% is used directly as fuel; the rest indirectly because the manufacture of farm machinery, pesticides, fertilizers, animal feeds, etc. depends on oil.

Farming practices

Different farming practices aim to increase productivity by:

- maximizing the rate of photosynthesis and therefore the growth of crops
- reducing losses in productivity because of
  - weeds which compete with crops for space and nutrients
  - animals (mainly insects) which eat crops and spread plant disease
  - fungi which cause plant diseases
  - the dissipation of energy raising livestock

Pesticides

Pests are organisms that reduce productivity by destroying crops and harming livestock. Pesticides are substances that kill pests and therefore help to increase productivity.

- *Insecticides* kill insects.
- *Herbicides* kill weeds.
- *Fungicides* kill fungi.

Pesticides are applied to crops as sprays, fogs, or granules and to livestock as dusts or dips.

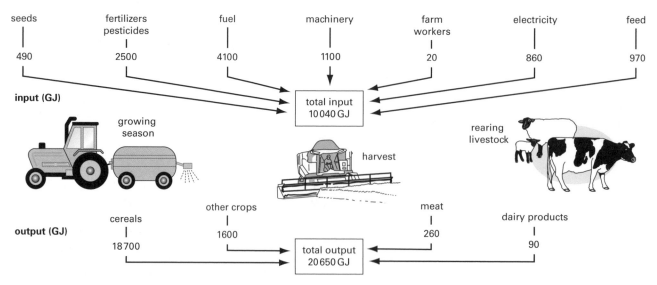

Energy inputs and outputs on a modern intensive farm; 1 GJ (gigajoules) = 10^3 MJ (megajoules) = 10^9 J (joules)

Fertilizers

The growth (and therefore productivity) of crops depends on elements and compounds that occur naturally in soil. Substances which add nutrients to soil are called **fertilizers**. They help to increase productivity by replacing the essential elements that crops take from the soil during the growing season.

The table lists the ions of some of the elements that crops need in relatively large amounts.

The ions of other elements are needed by plants in much smaller amounts (measured in tens of ppm or less). Many of them act as enzyme co-factors.

Element	ppm*	Ion	% of crop dry mass	Requirement
nitrogen	15 000	NO_3^-	3.5	synthesis of amino acids, proteins, and nucleic acids
potassium	10 000	K^+	3.4	enzyme co-factor, opening of stomata
calcium	5000	Ca^{2+}	0.7	formation of the plant cell wall
phosphorus	2000	PO_4^{3-}	0.4	synthesis of ATP and nucleic acids
magnesium	2000	Mg^{2+}	0.1	synthesis of chlorophyll
sulfur	1000	SO_4^{2-}	0.1	synthesis of some amino acids

*ppm = parts per million in solution

- **Natural fertilizers** (organic material such as manure and compost) are spread on soil. They help to maintain its structure. Fungi and bacteria decompose the material releasing nutrients which are absorbed by crops.
- **Artificial fertilizers** are added to soil as sprays or granules. Most supply nitrogen (N), phosphorus (P) and potassium (K) – the so called **NPK** fertilizers.

Rearing livestock

Livestock raised intensively are usually kept indoors. The aim is to reduce the dissipation of energy as heat from the animals so that they grow more quickly. The heat is released during cellular respiration.

- Confinement of animals in pens or cages restricts their movement.
  - As a result cellular respiration during muscle contraction is reduced.
  - As a result the energy dissipated in exercise is reduced.
- Their environment (heating, lighting) is controlled.
  - As a result the temperature difference between the environment and animals bodies is reduced.
  - As a result the energy dissipated in their keeping warm is reduced.

Rearing livestock intensively is sometimes called **factory farming**. The animals gain weight more quickly than those allowed to roam outdoors free range. Productivity therefore increases.

Costs and welfare

The ethical issues arising from modern intensive farming include

- costs to the environment
- the welfare of livestock

Fact file

In simple terms, estimates suggest that the practices of modern intensive farming increase productivity at a non-sun energy cost each year equivalent to more than 11 tonnes of oil for every person involved in the industry. The costs to the environment arise from the practices which increase productivity and which are the result of oil inputs.

- the use of pesticides which are poisonous and kill wildlife as well as pests – they may be a hazard to human health.
- the use of fertilizers which drain from land into water causing **eutrophication** – they may also be a hazard to human health.
- the use of modern farm machinery which works most efficiently in large open fields.
  - As a result the farming landscape is cleared of hedgerows, copses and woods.
  - As a result the biodiversity of ecosystems is reduced and genetic diversity lost.

Is it fair for livestock to be reared intensively so that we can enjoy eating more meat? Confining animals indoors causes them physical discomfort, boredom, and frustration. Some people argue this is cruel. Others claim that animals raised indoors must be content because they are safe from predators, sheltered, and eat well. However, we know that overeating in humans is a common sign of depression.

Questions

1 Use the data to calculate the energy value of the sun's input for a farm of 640 hectares.

Energy received at ground level = 0.35 MJ cm^{-2} yr^{-1} (100 million cm^2 = 1 hectare).

2 Using the answer from question 1 with the following data, calculate the amount of light energy converted into crop biomass each year.
- Photosynthetic efficiency = 0.8%
- Average crop cover = 50%
- Length of time crops cover the soil each year = 6 months

3 For every 1 m^2 of grass it eats, a cow obtains 3000 kJ of energy. It uses 100 kJ in growth, 1000 kJ are lost as body heat, and 1900 kJ are lost in faeces.

a What percentage of the energy in 1 m^2 of grass
  (i) is used in growth, and
  (ii) passes through gut as food which is not absorbed?

b If beef has an energy value of 12 kJ g^{-1}, how many m^2 of grass are needed to produce 100 g of beef?

2.26 Succession

Communities do not stay the same forever. They change over time as one community gives way to another. The process is called **succession**. The communities of a succession are each called a **sere**. A sere can be recognized by the collection of species that dominate at that point in a succession.

A succession begins with the colonization of a new environment which is clear of organisms because of a **disturbance**. The colonizers are called **pioneers** and form a **pioneer community** of producers and consumers. The final sere of a succession is called the **climax community**, which is stable. Its species make up does not change over time unless a disturbance removes them. The idea is the same for any succession. Here the diagram shows the succession of a pond as an example.

Notice:

- The different species of the pond community forming a sere impact on their own environment, altering it. The changes may favour the colonization of new species rather than the survival of the original species causing the changes.
  - As a result the previously dominant species may die out to be replaced by the new species.
- As the succession develops, the community of each sere is more **biodiverse** (made up of more species) than the community of the previous sere.
  - As a result the flow of energy through the community increases from one sere to the next.
  - As a result the productivity of the ecosystem increases as the succession develops.
  - As a result the biomass of the community of a sere is greater than the biomass of the community of the sere preceding it.
- Succession is directional. The different serial stages in a particular succession (e.g. pond) can be predicted.

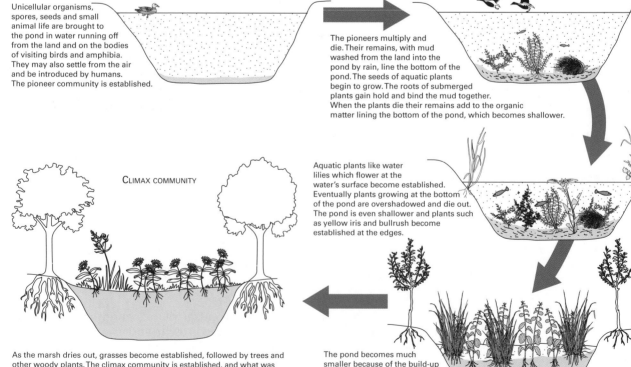

PIONEER COMMUNITY

Unicellular organisms, spores, seeds and small animal life are brought to the pond in water running off from the land and on the bodies of visiting birds and amphibia. They may also settle from the air and be introduced by humans. The pioneer community is established.

The pioneers multiply and die. Their remains, with mud washed from the land into the pond by rain, line the bottom of the pond. The seeds of aquatic plants begin to grow. The roots of submerged plants gain hold and bind the mud together. When the plants die their remains add to the organic matter lining the bottom of the pond, which becomes shallower.

Aquatic plants like water lilies which flower at the water's surface become established. Eventually plants growing at the bottom of the pond are overshadowed and die out. The pond is even shallower and plants such as yellow iris and bullrush become established at the edges.

CLIMAX COMMUNITY

As the marsh dries out, grasses become established, followed by trees and other woody plants. The climax community is established, and what was once a pond is now dry land. The succession may take several hundred years to complete.

The pond becomes much smaller because of the build-up of mud and organic material. Eventually it becomes a marsh.

The serial stages in the succession of a pond

Primary and secondary succession

Succession which begins when organisms colonize an environment where previously living things were absent is called **primary succession**.

- Islands formed from volcanic eruptions (the disturbance) undersea are examples of environments where the organisms of a pioneer community establish a primary succession on the bare rock.

- Lichens are one of the few types of organism able to survive in such hostile conditions. Their activities break down the rock into particles. Their dead remains decompose, adding nutrients to the mixture, forming soil in which the species of the next sere can establish a foothold.

Succession which begins when organisms colonize an environment where previously living things were established is called a **secondary succession**.

- Land cleared for agriculture or forest destroyed by fire (the disturbances) are examples of environments bare of wildlife. However, recolonization quickly takes places if circumstances allow. Surviving seeds, spores, and the parts of plants buried underground and capable of asexual reproduction are the starting point for a new succession. In time the climax community is re-established if the succession is undisturbed.

- Succession occurs on different time scales, ranging from a few days to hundreds of years. For example, the development of a climax woodland may take hundreds of years; the succession of insects and fungi in a pat of cow dung may take just a few months.

Deflected succession

Some communities seem to be stable and remain unchanged for long periods of time. Their stability, however, is the result of human activities. For example, we think of moorland as 'natural'. However, the moorland environment is the result of clearance of the climax community of woodland to allow grazing (the disturbance) by livestock.

In other words moorland is sub-climatic, but persists because long-term **grazing** deflects the succession from its climax. Grazing destroys seedling trees, so that the climax community of woodland cannot develop. We call moorland a **plagioclimax**.

Moorland has amenity value (we enjoy the landscape for its beauty and leisure opportunities). Our understanding of how to manage its succession through grazing by livestock long-term underpins the **conservation** of a highly prized environment.

A mowed lawn is another example of deflected succession resulting in a plagioclimax – providing that mowing (the disturbance) occurs regularly. However, if the lawn is left uncut daisies and dandelions soon appear, followed by a range of other flowering plants as succession gets underway. If the lawn is left for a few years, woody plants grow up and overshadow the grass, which dies out.

Questions

1. What is the difference between a primary succession and a secondary succession?

2. Briefly describe the relationship between flow of energy through and productivity of the communities of a succession from pioneer to climax.

3. What is a plagioclimax?

Populations and the ecosystems where they live are usually too large for it to be practical to study everything about them. Instead **samples** are taken of the ecosystem under investigation. The samples are assumed to be representative of the ecosystem as a whole. In order to study an ecosystem, the different species that live in it are quantified using two factors:

- the **distribution** – where the organisms are in the ecosystem
- the **abundance** – the number of individuals of a species relative to other species in the ecosystem

These quantities are measured using sampling methods such as line transects, belt transects, quadrats, and point quadrats.

Fieldwork produces data which makes it possible to draw conclusions about populations and ecosystems in general. The diagram illustrates different ways of sampling populations in the different habitats of a woodland

ecosystem. Using equipment which is suitable for sampling the populations is an important choice if the fieldwork is to be successful. For example, quadrats would not be useful to sample populations of fast moving animals.

An investigation is designed so that samples are taken at **random**. Random sampling means that any part of an ecosystem (its abiotic environment/populations) has an equal chance of being sampled. Different methods are used to ensure random sampling, such as using tables of random numbers to select samples.

Random sampling avoids bias in the data gathered during an investigation. Biased data might lead to false conclusions. We can never be absolutely sure of any conclusions drawn from data. The conclusions are **tentative** (not definitive). However we can improve confidence in data by reducing error in the sampling methods used to obtain the data.

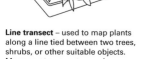

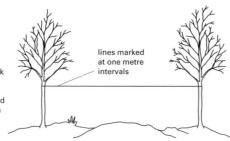

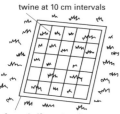

Tree beating – a stick is used to tap the branches of shrubs or trees, knocking small animals living there on to a tray or white cloth.
- do not damage the shrub or tree
- count the animals and identify them

Reducing sampling errors
- take enough samples
- take samples following a *consistent* pattern (e.g. tap branches the same number of times for each sample)
- do not sample the same branch more than once

tray

Line transect – used to map plants along a line tied between two trees, shrubs, or other suitable objects. Measure at one metre marks.
- distance to the ground
- plant species growing under the mark
- height of plant(s)

Reducing sampling errors – only record the plants growing directly under each of the one metre marks along the line of transect

lines marked at one metre intervals

twine at 10 cm intervals

frame half a metre along each side – called a quadrat

Quadrat – a square frame used to identify and count the number of plants or animals in a *known* area.
- throw the quadrat at random in the study area
- count the plants or animals and identify them
- calculate abundance as the number of squares that the plant or animal occupies

Point quadrat – a horizontal wooden or metal frame with pointed needles pushed through at regular intervals. Each plant touched by the needle is recorded

Sweep net – a strong cotton bag attached to a handle and swept through tall grass and other vegetation. 'Swishing' the net through the grass knocks small animals (e.g. insects, spiders) into the bag.
- count the animals and identify them

Reducing sampling errors:
- take a sufficient number of samples
- take samples following a *consistent* pattern (e.g. walk ten paces while 'swishing', taking one 'swish' for each step)
- do not sample the same area of vegetation more than once

Belt transect – uses quadrat and line transect together. Useful for measuring changes in vegetation between two points.
- lay down tape or rope along the line of the transect
- use a quadrat to record the plant species at intervals along the transect
- estimate abundance of each species

Reducing sampling errors
- the position of the quadrat along the line of transect should follow a *consistent* pattern

quadrat — line of transect

positions of quadrat along transect

Quadrats are not suitable for sampling mobile animals. However they are suitable for sampling small static animals (e.g. limpets clinging to the rocks by the sea shore)

Fieldwork describes what is done to gather data from ecosystems

Reducing sampling error

In general, ways of reducing error include

- random sampling
- consistent methods of working
- taking a number of samples (the more the better – within practical limits!)
- standardizing when samples are taken (e.g. at the same time of day, same season, similar weather conditions)

… all contribute to methods of field work which enable reliable data to be gathered for analysis. Data analysis leads to conclusions about the populations and ecosystems under investigation. Appropriate statistical tests of the data help to improve confidence in the conclusions drawn.

Ethical issues

Fieldwork in the tree tops of tropical rainforest is a challenge. However, different methods of sampling the populations living there help scientists to collect data which is improving our understanding of biodiversity (and genetic diversity). For example, more than 30 million species of insect new to science are estimated to be living in the branches, leaves, fruits and flowers of rainforest trees – and insects are only one taxon (group) of the animal kingdom.

Methods of sampling rainforest insects fall into two general categories:

- bringing the insects to the ground where scientists collect them

- enabling scientists to reach the insects in the tree tops

For example:

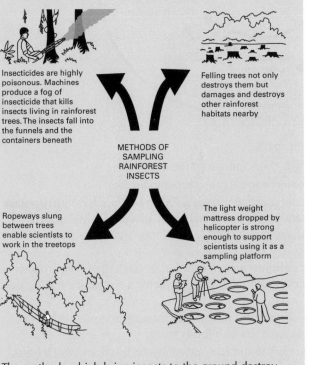

Insecticides are highly poisonous. Machines produce a fog of insecticide that kills insects living in rainforest trees. The insects fall into the funnels and the containers beneath

Felling trees not only destroys them but damages and destroys other rainforest habitats nearby

METHODS OF SAMPLING RAINFOREST INSECTS

Ropeways slung between trees enable scientists to work in the treetops

The light weight mattress dropped by helicopter is strong enough to support scientists using it as a sampling platform

The methods which bring insects to the ground destroy populations and habitats. Scientists working in the tree tops disturb the habitats of animals living there. The ethical issues arising from the fieldwork might determine which methods are used to sample rainforest insects.

Where do you think the balance lies? Make a list of *fors* and *againsts* for each of the methods of sampling rainforest insects shown in the diagram.

Estimating population size

Only by counting all the individuals of a population can an absolute figure be placed on the size of a population. This approach is only possible for individuals of **sessile** (static) species which are spread out and easily seen. For mobile species (individuals moving from place to place) different ways of estimating population size are used. One method is called the **capture/recapture** method or the **Lincoln Index** (named after the American biologist who worked it out).

How does the method work?

Imagine you want to find the size of a population of a species of snail. You need a small quantity of dull, water-based paint and a thin paintbrush (a twig will do).

- Dab a small spot of paint onto a known number of snails (not less than 40), returning each one to the spot where you found it.
- After a day or so return and collect a similar number of snails. Do not specially look out for marked snails.
- Count the number of marked snails in your second collection.

The size of the snail population is estimated as follows:

$$\text{population size} = \frac{\text{number of snails marked first time} \times \text{number of snails collected second time}}{\text{number of marked snails collected second time}}$$

Estimating the size of a snail population using the Lincoln Index

A person collects a sample of 84 snails of a particular species from a small area of grass covered bank, marks them with a dab of paint and returns them to the place from where they were collected. A few days later the person collects a sample of 85 snails from the same area. 42 snails of this second sample are marked with a dab of paint.

Q Estimate the size of the snail population.

A *Using the Lincoln Index:*

$$\text{the size of the snail population} = \frac{84 \times 85}{42} = 170$$

Q Why is dull, water-based paint used to mark the snails?

A *Dull paint decreases the risk of marked snails being noticed by predators. Water-based paint eventually washes off and does not damage the snail.*

Q Why should you not specially look out for marked snails?

A *If marked snails are deliberately collected then the size of the snail population will be under-estimated.*

Questions

1 Froghoppers (small insects) live in long grass. A sample of 90 froghoppers was collected in a field using a sweep net. Each froghopper was marked with a spot of dull water soluble paint. A second sample taken 24 hours later produced a sample of 80 froghoppers. Six were marked.

   a Estimate the population of froghoppers in the field.

   b Explain why dull water soluble paint was used to mark the froghoppers.

The processes of decomposition release mineral nutrients and gases into the environment.

The nitrogen cycle

The concept map and its checklist are your revision guide to the nitrogen cycle.

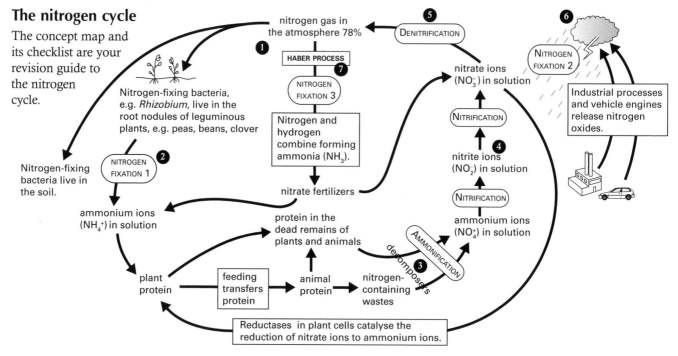

Fact file

Nitrate ions absorbed by plant cells are reduced to nitrite ions and then ammonium ions before becoming part of nucleic acids, proteins, and other nitrogen containing compounds. The reduction reactions are catalysed by **reductase** enzymes.

Fact file

The conversion of ammonium compounds into nitrates (nitrification) releases energy which is then available to drive the metabolism of the bacteria responsible. The reactions are an example of **chemosynthesis**. The bacteria are referred to as **chemoautotrophs**.

Questions

1 Outline the stages in the nitrogen cycle.

2 Why are fields sown with leguminous crops (e.g. peas and beans) in 'rotation' with other crops?

Checklist: the nitrogen cycle

❶ • Nitrogen is the most abundant gas in the atmosphere.
  • Most organisms *cannot* use gaseous nitrogen directly.

❷ • Some types of bacteria called **nitrogen fixing** bacteria have **nitrogenase** enzymes which catalyse the combination of gaseous nitrogen with hydrogen, forming ammonia (NH_3). In solution ammonia forms ammonium ions (NH_4^+).
  • Plants take up ammonium ions in solution through their roots (plants also take up nitrate ions in solution) enabling them to synthesize nucleic acids and proteins.
  • Feeding transfers nitrogen (as nitrogen containing compounds) from producers → consumers through food chains.

❸ • **Ammonification** refers to the enzyme-catalysed reactions which break down dead organisms and nitrogen containing wastes. A variety of decomposers are responsible.
  ☞ As a result ammonium ions (NH_4^+) are formed.

❹ • **Nitrification** refers to the oxidation reactions carried out by **nitrifying bacteria**. The reactions convert ammonium (NH_4^+) compounds to nitrites (NO_2^-, e.g. by *Nitrosomonas*) and then to nitrates (NO_3^-, e.g. by *Nitrobacter*).
  ☞ As a result nitrogen as nitrates (NO_3^-) is available in a form that is most easily absorbed by the roots of plants.

❺ • **Dentrification** refers to the reduction reactions carried out by **dentrifying bacteria**. The reactions convert nitrate ions (NO_3^-) to nitrogen gas (N_2).
  ☞ As a result nitrogen gas enters the atmosphere.

❻ • Lightning is a high energy discharge which breaks apart nitrogen molecules. The nitrogen atoms produced react with atmospheric oxygen.
  ☞ As a result nitrogen oxides form, which dissolve in rain drops. Nitric acid (HNO_3) and nitrous acid (HNO_2) are produced.
  • The acids react with compounds in the soil, forming nitrates (NO_3^-) and nitrites (NO_2^-).

❼ • The Haber reaction is an industrial process which fixes nitrogen in combination with hydrogen producing ammonia (NH_3).
  • Some of the ammonia is used to make nitrate fertilizers.
  ☞ As a result crop production is improved.

Biotic factors affect population size

Imagine a pioneer species colonizing a new area. Provided there are no shortages of food and few predators individuals are likely to survive and reproduce. Their numbers increase, and the population grows. The graph shows you how. It is called a **sigmoid growth curve** because it is S-shaped: sigma is the Greek for 's'. Follow the sequence of numbers on the graph.

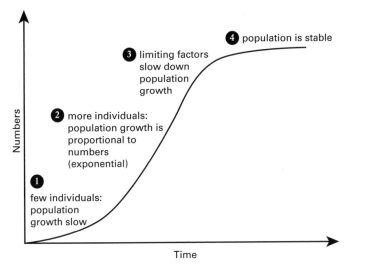

❹ population is stable

❸ limiting factors slow down population growth

❷ more individuals: population growth is proportional to numbers (exponential)

❶ few individuals: population growth slow

Numbers

Time

❶ To begin with the growth rate (number of individuals added to the population in a given time) is slow. This is called the **lag phase**. *Why?* There are only a few individuals available for reproduction.

❷ As the number of individuals increases, the growth rate increases. This is called the **exponential phase**.

Why? More individuals are available to reproduce. However, there is little intraspecific competition for food and the effects of disease and predators are slight. Notice that the growth curve becomes steeper. At this stage the rate of increase in the population is proportional to the numbers of individuals present.

   As a result the numbers of each generation are double the numbers of the previous generation. We say that population growth is geometric because the numbers of individuals of successive generations form a geometric series (e.g. 100, 200, 400, 800, 1600…)

❸ The geometric growth of a population does not continue indefinitely. As numbers build up, shortage of food, the effects of disease and predators, and other **limiting factors** such as lack of shelter and the accumulation of poisonous wastes slow the rate of growth. This is called the **phase of deceleration**.

Why? Limiting factors increase the **environmental resistance** to further population growth. The effects are **density dependent** – the more individuals there are in a particular area the greater is intraspecific competition for food, and the more likely it is that individuals will fall victims of predators and disease.

❹ Eventually, the rate of population growth levels off. This is called the **phase of stability**.

Why? The factors that add individuals to a population (births and immigration) are balanced by those that remove individuals from a population (deaths and emigration). In reality the numbers of the population fluctuate around a mean value which represents the maximum size of the population. The term **carrying capacity** refers to the level of resources of an environment able to sustain the maximum numbers of a population.

Remember that

- **populations** are each made up of a group of individuals of a particular species living in a particular place at a particular time

- **births** and **immigration** increase the size of a population

- **deaths** and **emigration** decrease the size of a population

Fact file

The density of a population is the number of individuals living in a given area (land) or volume (water) of habitat.

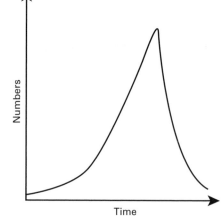

Abiotic factors affect population size

Population growth does not always follow the pattern of a sigmoid curve. For some species, populations may rapidly increase only to 'crash'. The growth curve is **J-shaped** (see left).

The steep decline in numbers is usually the result of a sudden change in abiotic conditions, for example a sharp drop in temperature. The effect is **density independent** – the 'crash' occurs regardless of the size of the population and before limiting factors begin to reduce its rate of growth.

Predator and prey populations

Predators affect the numbers of their prey. Prey has an effect on the numbers of predators. If prey is scarce then some predators starve. The graph below shows the relationship between the numbers of a predator and the numbers of its prey. Follow the sequence of numbers.

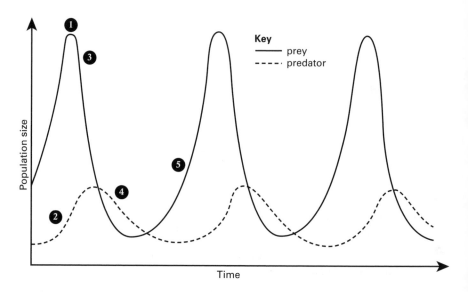

① Prey breed and increase in numbers if conditions are favourable (e.g. food is abundant).

② Predators breed and increase in numbers in response to the abundance of prey.

③ Predation pressure increases and the number of prey declines.

④ Predator numbers decline in response to the shortage of food.

⑤ Predation pressure decreases and so prey numbers increase… and so on.

Notice that

• fluctuations in predator numbers are less than fluctuations in prey numbers

• fluctuations in predator numbers lag behind fluctuations in prey numbers

Why is this? There are fewer predators than prey, and predators tend to reproduce more slowly than prey.

Notice also that changes in the numbers of each population are density dependent.

• When the numbers of prey increase so too do the numbers of its predator, until the deaths of prey from predation exceed the number of new individuals entering the prey population through births.

    Ⓡ As a result the numbers of prey decrease, followed by a decrease in the number of predators, because of the lack of food. The effect is density dependent and an example of negative feedback returning a change of state (in this case, numbers of predator and prey) to a normal value (homeostasis).

Questions

1 How does a sigmoid growth curve of a population describe the increase in its numbers?

2 Give the meaning of the phrase 'density dependent'.

3 Explain why fluctuations in numbers of predator lag behind that of its prey.

Adaptations to abiotic factors: an example

Why do woodlice live under stones, in layers of detritus and other dark, damp places?

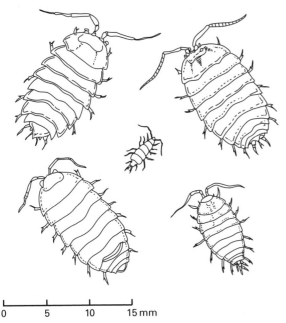

European woodlice – different species are different sizes

- *Remember* that woodlice are arthropods. The exoskeleton that encases the woodlouse body lacks the waxy waterproof layer which impregnates the exoskeleton of other arthropods such as insects and spiders.
  - Ⓡ As a result woodlice quickly lose water from the body in dry air, threatening their survival.
  - Ⓡ As a result woodlice live in dark, damp places where the air is saturated with water vapour (**humid**).
  - Ⓡ As a result loss of water from the woodlouse body is reduced.
- Damp places are an example of a **microhabitat**. They provide highly localized environments in miniature where abiotic factors may be different from the abiotic factors of the habitat as a whole.
  - Ⓡ As a result woodlice can survive where they would perish elsewhere in the habitat.

Abiotic factors affect the distribution of organisms

The term 'distribution of organisms' refers to where living things are found in an ecosystem. Abiotic factors affect the distribution of organisms. They include:

Light
The intensity of light is an important abiotic influence on the distribution of organisms inside temperate woodland. It affects the rate of photosynthesis and therefore the amount of plant growth under the canopy layer of the trees.

- Plants of the field layer grow and flower in the spring. The plants take advantage of the increasing intensity of sunlight for photosynthesis. The sunlight reaches the woodland floor through branches bare of leaves.

- Ⓡ As a result, more food and shelter are available for animals.

Rain and temperature
With rain and warmth, communities flourish. The diagram shows the effect of seasonal rainfall on the distribution of plants in West Africa. Weather stations are located at the places named on the map. Changes in the type of vegetation are shown along a transect marked A–B.

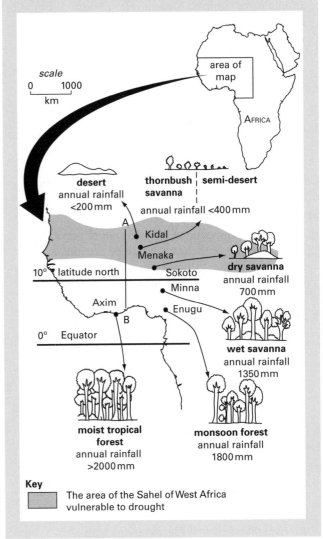

Biotic factors affect the distribution of organisms

The biotic factors which affect the distribution of organisms include competition, predator/prey interactions, and other interactions.

Competition

In nature, living things are **competitors** (rivals) for resources which are *in limited supply*. The resources include water, food, light, space, and mates.

- **Intraspecific competition** refers to competition between individuals of the same species. For example cacti are widely spaced apart. They look as if they have been planted out in a regular arrangement. Although many tiny cactus seedlings sprout in a particular area, the pattern appears because there is only enough water for some of them to grow into mature plants. Growing cacti are the competitors and water is the resource in short supply. In particular, those cacti whose root systems develop and spread the most extensively underground absorb the most water, depriving their slower-growing rivals. The extent of each root system determines the distance between neighbouring cacti.

- **Interspecific competition** refers to competition between individuals of different species. It often leads to one species displacing another from a particular niche. For example, red squirrels (*Sciurus vulgaris*) were common in woodlands throughout Britain before the introduction of grey squirrels (*S. carolinensis*) from North America in the 1870s. Now, red squirrels are restricted to a few pockets of woodland in England although they remain fairly common in Scotland. It seems that both species compete for similar types of food but that grey squirrels are more efficient at the task. In Scotland, grey squirrels may be at the northern limits of their climate range and find it more difficult than red squirrels to survive the harsher weather.

Competition between organisms not only affects their distribution. It also has other outcomes.

Short-term outcomes:

- Intraspecific competition helps to control population size. For example the competition by cacti for water, means that slower growing plants are more likely to be affected by water shortage and therefore less likely to survive.

- Interspecific competition ensures that only one species occupies a particular niche at a particular time (the Gause hypothesis).
  - As a result available space, food, water, etc. is shared between different species.
  - As a result the number of species occupying a habitat is at a maximum (biodiversity) with respect to the environmental resources available.

Long-term outcomes:

- Intraspecific competition is a component of **natural selection**. The individuals whose characteristics best adapt them to obtain environmental resources are more likely to reproduce (because they are more likely to survive) than less well adapted individuals.

- As a result their offspring will inherit the genes responsible for the favourable characteristics.
- As a result favourable genes accumulate in the population and the species changes (**evolves**) over time.

- Interspecific competition leads to **competitive exclusion** (a species excludes another species from a particular niche).
  - As a result the displaced species becomes **extinct** unless it is able to transfer and adapt to another niche.

Predator/prey interactions

Predators attempt to catch prey. Prey attempt to escape predators. The interaction between predator and prey is intense: not least because the predator requires food and predation is fatal for the prey! The relationship affects the distribution of predators and prey. Put simply, predators gather where prey is abundant and prey will tend to avoid predators.

Other interactions

'Interaction' means the way in which a species affects another species including their respective distributions. Competition and the relationship between predator and prey are examples. Other forms of interaction include:

- **mutualism** – two or more species benefit from their close relationship. Pollination in flowering plants often depends on insects. In return, the flower's visitors benefit from the sweet-tasting nectar stored in the nectaries.

- **commensalism** – an association between one species that benefits (the commensal) and another which is unaffected either way. For example some species of fish are immune to the sting cells on the tentacles of sea anemones. Living among the tentacles means that the fish gain protection from predators. The sea anemone seems to be unaffected by the fish.

- **parasitism** – an association between one species that benefits (parasite) and another which is harmed (host). The parasite is usually much smaller than the host. Beef tapeworms live in the human intestines surrounded by digested food. They absorb food through the body wall. The host is deprived of food and the intestine may become blocked. The tapeworm's wastes cause illness.

Questions

1 What are abiotic factors?

2 Explain the difference between intraspecific competition and interspecific competition.

3 Explain the difference between an organism's habitat and its niche.

2.31 Conservation

Conservation of species

There are several possible strategies:

- **Preservation** involves keeping some part of the environment without any change. This is only possible if the area can be fenced off or otherwise protected from human exploitation.

Conservation methods are typically more dynamic:

- **Management** involves interventions to actively conserve the ecosystem, and/or to use its resources in a sustainable way.
- **Reclamation** involves the restoration of damaged habitats. This often applies to the recovery of former industrial sites such as mineworkings.

Managing ecosystems to provide resources

Detailed knowledge about an ecosystem allows humans to manage it in such a way that it can provide resources in a sustainable way. For example, timber can be produced in temperate countries such as the UK by sustainable forestry. Methods include cutting down only selected mature trees, or selected areas of an established forest. Unlike clearing the entire forest, this leaves areas to sustain the biodiversity of the ecosystem.

In **coppicing** tree species within a forest or woodland are cut down close to the ground and the new shoots that grow are cut and used periodically. A coppiced woodland often contains several different species in sections that are harvested in rotation so that a crop is ready each year. Some trees called **standards** are not coppiced but are allowed to grow to maturity. A coppiced wood provides a rich variety of habitats as there is a range of coppices of different ages. Different species are cut at different times depending on the local custom, and the intended use for the timber. Birch can be coppiced for brushwood on a 3–4-year cycle, whereas oak can be harvested over a 50-year cycle for poles or firewood. Other species that are coppiced include hazel, hornbeam, beech, ash, alder, and willow.

Another sustainable method of forestry similar to coppicing is **pollarding** in which the upper branches of a tree are removed, resulting in the growth of a dense head of foliage and branches. Pollarding may be preferred to coppicing in grazed areas, because animals browse the regrowth from coppice stools. Pollarding in woodland encourages underbrush growth due to increased levels of light reaching the woodland floor. This can increase species diversity.

Reasons for conservation

It is in the interests of the human race to preserve the biodiversity of the Earth – both plant and animal species – for many reasons, including:

- Economic reasons: biological resources are useful to us as they provide us with food, drugs, and products such as timber, dyes, and oils.
- Ecological reasons: the organisms in an ecosystem interact with each other in complex ways and if one species is lost this can upset the natural balance and have unforeseen consequences for the rest of the ecosystem.
- Ethical reasons: humans are a dominant and powerful species. We have a duty not to knowingly destroy habitats or species but to conserve them for future generations.
- Aesthetic reasons: the beauty and variety of the many diverse ecosystems on Earth provide pleasure for many people as seen by the importance of travel and tourism in undeveloped regions.

Conservation issues in the Galapagos Islands

Humans cause problems!

- Ecotourism – generates money **but** increases pollution

- Increasing conversion of land to agricultural use removes 'wild' habitat and diverts fresh water

- Use of wild species as a food resource (the Beagle took more than 100 tortoises!)

- Oil spillage – disastrous for marine iguanas and seals

- Overfishing – nets also drown dolphins, iguanas and penguins.

But humans are conservationists too!

- removal of introduced species e.g. Isabela is now free of goats

- reintroduction of species – Espanola tortoises (there are plenty) have been moved to Pinta (where there's only one left!)

- increasing landing fees to limit tourism.

Fernandina

Santiago

Isabela

Chavez

San Cristobal

Santa Fe

El Nino:

- follows warm ocean conditions off the west coast of Ecuador and Peru

- changes to sea temperatures and wind patterns affect rainfall – becomes very heavy in these regions

- rising sea temperatures reduce nutrient content of water, affecting ocean food chains.

Galapagos: ideas on natural selection

Charles Darwin, on the voyage of the Beagle, made many observations that contributed to his ideas on Natural Selection. These included variations in the islands' tortoises and mockingbirds.

Later research has shown that finches of the Galapagos islands show adaptations to the food resources on the different islands.

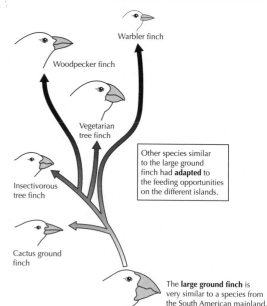

Warbler finch

Woodpecker finch

Vegetarian tree finch

Insectivorous tree finch

Other species similar to the large ground finch had **adapted** to the feeding opportunities on the different islands.

Cactus ground finch

The **large ground finch** is very similar to a species from the South American mainland.

Alien invaders!

Introduced species arrived when sailors (and pirates) visited the islands for food and fresh water.

Goats: eat all vegetation – reduce food available to tortoises and land iguanas.

Pigs: omnivorous scavengers which also destroy delicate ecosystems as they dig for plant roots.

Rats: a particular problem as they eat eggs of ground-nesting birds (e.g. boobies).

Cats: direct predators on ground-nesting species.

A flagship species: attractive, so attracts financial support used for benefit of whole habitat.

Galapagos Penguin

Endangered by:

- El Nino

- Overfishing of prey species

- Feral cats

- Becoming trapped in fishing nets.

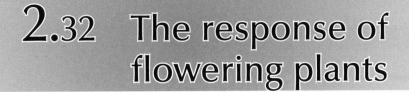

Plants need to respond to their environment to avoid predation and abiotic stress. *Recall* that in flowering plants, many of their responses to stimuli are movements which are the result of growth. Stimuli which are more intense from one direction are called **directional stimuli**. For example, light shining from a particular direction (unidirectional) stimulates shoots to bend and grow towards it. Growth movements in response to directional stimuli are called **tropisms**.

Discovering indoleacetic acid (IAA)

In 1928, the Dutch plant biologist Frits Went investigated the response of seedlings to unidirectional light. The results of his experiments suggested that a substance
- is produced in the tips of coleoptiles
- passes to the region behind the tip
- stimulates growth so that the coleoptiles grow towards unidirectional light.

Went found that the extent to which coleoptiles grow towards directional light is proportional to the amount of substance present. His observation is an example of biological measurement as a result of **bioassay**. The technique allows comparison of the effect of an unknown amount of a substance with the effect that samples of known amounts of the substance have on a biological system (in this case the angle of bending of coleoptiles to directional light). The diagram shows you the idea.

The bending of the coleoptile suggests that the increase in concentration of the substance in the tissues of one side of the coleoptile promotes growth on that side. As a result the coleoptile bends over.

1 A coleoptile is decapitated and the tip placed on top of a block of agar.

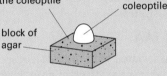

substance absorbed by an agar block from the tip of the coleoptile

tip of coleoptile

block of agar

2 The block of agar is placed on one side of the decapitated coleoptile.

decapitated coleoptile

the substance passes from the agar block to the tissues of one side of the decapitated coleoptile.

3 The decapitated coleoptile bends over. The angle at which it bends is measured.

angle

The substance that Went collected was a plant hormone called **auxin** (from the Greek word auxein, 'to grow'). In 1934, auxin was identified as **indoleacetic acid (IAA)**. It is widely distributed in plant tissues and regulates their growth, by promoting cell elongation.

Phototropism, geotropism, and auxin

IAA is synthesized in young leaves and the tips of shoots. A small amount may also be synthesized in the tips of roots. Its transport is by
- diffusion from cell to cell
- translocation in the phloem from shoots to roots

Elongation of cells is either *stimulated* or *inhibited* depending on the concentration of IAA and its location in plant tissues. The graph and diagram on the next page show you the idea.

Fact file

The coleoptile is a sheath of tissue which encloses the shoot of a germinating grass seedling. It grows like a shoot. Its uncluttered structure makes its growth easier to observe.

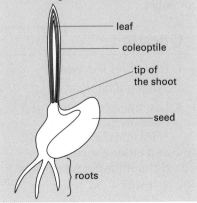

leaf

coleoptile

tip of the shoot

seed

roots

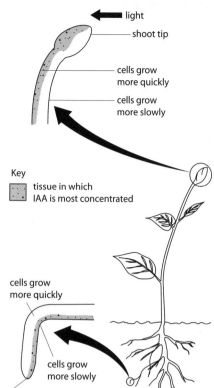

IAA stimulates elongation in shoots and inhibits it in roots

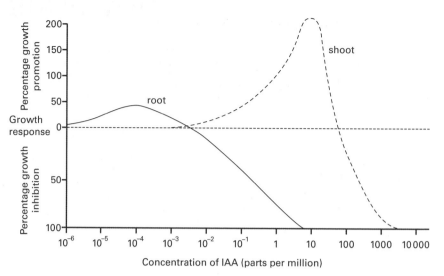

Growth of shoot and root in response to the concentration of IAA. The scale for IAA concentration is logarithmic

Notice that the concentration of IAA which stimulates the growth of shoots also inhibits the growth of roots. The diagram illustrates the outcome of the different responses.

Research shows that, in seedlings, auxin accumulates on the shaded side of a shoot tip illuminated by directional light. The faster elongation of cells on the shaded side compared with the brightly lit side results in curvature of the shoot towards the light source.

In the root tip of seedlings, accumulation of auxin on the lower side produces an opposite response. The elongation of cells is inhibited while the cells on the upper side elongate faster, with the result that the root grows downwards. Downward growth occurs even if the root of the seedling is exposed to light. In other words, roots are negatively phototropic and positively geotropic as the diagram above illustrates.

Other plant hormones and their interactions

After IAA was isolated other plant hormones were soon discovered. Today we know that a range of substances regulate different aspects of plant growth. The substances include **gibberellins**, **cytokinins**, **abscisic acid**, and **ethene**.

The responses of plants to stimuli are often the result of the interactions between the different hormones. If one hormone enhances the effect of another then the interaction is stimulatory. We say it is **synergistic**. If the effect of a hormone is reduced by another, then the interaction is inhibitory. We say it is **antagonistic**. Overall plant responses are the result of the balance between synergistic and antagonistic effects of the plant hormones.

Other roles of IAA

IAA is not only associated with the tropic responses of plants. It is also responsible for other aspects of plant growth. *Remember* that the action of IAA as a promoter or inhibitor of growth depends on its concentration and the tissue in question.

- **Apical dominance** – IAA present in the tip of the shoot apex (the top of the shoot) inhibits cell division in the lateral (side) buds below, preventing the growth of lateral shoots. The term apical dominance refers to the relationship between the apex and lateral buds of a shoot. Lopping of the shoot apex removes the source of IAA and side shoots develop. This is why a gardener trims a hedge to make it more bushy.

- **Leaf fall** – IAA is highly concentrated in young leaves. It inhibits their fall by preventing formation of the layer of tissue (called the **abscission layer**) which develops at the base of the stalk of older leaves before they fall from the plant. The abscission layer forms because the concentration of IAA in leaves declines as they age.

Gibberellins

Gibberellins are a separate group of plant hormones that stimulate the growth of stems and leaves. Like IAA it stimulates cell elongation. Evidence for the action of gibberellins:

- They stimulate rapid growth in plants with inherited dwarfism thought to be due to an inability to synthesize gibberellins

- Rice plants infected with a fungus called *Gibberella* contain excessive amounts of gibberellins. They grow tall and fall over before reaching maturity.

Questions

1 How does IAA work?

2 What is a bioassay?

3 Use the internet to find out how gibberellins, cytokinins, abscisic acid, and ethene regulate different aspects of plant growth.

Commercial uses of plant hormones

Auxins are used as defoliants, e.g. during the Vietnam War to clear areas of vegetation and make bombing of bridges, roads, and troops easier. Also used to remove vegetation from overhead power lines – manual removal would be costly and dangerous.

A mixture of **auxin, cytokinin** and **gibberellin** will inhibit apical growth and allow limited development of lateral buds. This mixture applied to hedges promotes dense, bushy growth and limits the need for mechanical trimming to one or two occasions per year.

Gibberellic acid may mimic red light: control of flowering time (promote long-day species, inhibit short-day species) means flowers can be available 'out of season'.

Ethene sprayed onto day-neutral species such as pineapple can synchronise flowering/fruiting so that crop picking can be more efficient.

Auxin can act as a selective lawn weed killer since broad leaved 'weed' species are killed by auxin concentrations which do not affect monocotyledons.

Cytokinins delay leaf senescence and are used to maintain the life of fresh, leafy crops such as lettuce.

Auxin can inhibit 'sprouting' (lateral bud development) in stored potatoes.

Plant hormones and fruit growing

Cytokinin can promote fruit growth.

Gibberellins increase fruit size in grapes since some ovules abort – 'crowding' is reduced allowing more nutrients to reach remaining fruits and limiting fungal infections.

N.B. Many commercial applications of these growth phenomena rely on **plant growth regulators,** which are synthetic derivatives of the natural compounds, but are usually more effective in lower concentrations because they are degraded less rapidly by the plant.

Auxin can prevent premature fruit drop (windfall losses) since it is antagonistic to **abscisic acid.**

Ethene is used to accelerate ripening – ideal for grapes which can be picked earlier and thus have a longer drying period for forming raisins. Ripening can be delayed by keeping fruits in an oxygen-free atmosphere: ethene can then induce ripening as required.

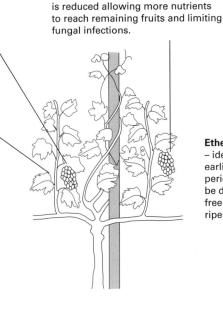

Animal responses

Sensitivity is an obvious life process for animals. Animals such as humans need to respond to stimuli in the environment around them in order to find food, find a mate, avoid dangerous substances or situations, interact socially, and for many other reasons. Human responses are brought about by a combination of hormonal, nervous, and muscular coordination.

How the nervous system is organized

Neurones are grouped into bundles called **nerves** which pass to all of the muscles and glands (the effectors) of the body. The nerves form the **nervous system**:

- the **central nervous system** consists of the brain and spinal (nerve) cord
- the **peripheral nervous system** consists of the cranial nerves and spinal nerves joining the central nervous system.

The diagram illustrates the arrangement.

If an individual's response to a stimulus is controlled by the brain it is called a **voluntary** response. Thinking and decision are part of the process. If thinking and decision are not involved then the response is said to be **involuntary**. A **reflex response** is involuntary.

The **autonomic nervous system** controls involuntary responses: for example, movement of the gut and the release of some of its internal secretions, breathing movements, and the beating of the heart. It forms part of the peripheral nervous system and consists of the **sympathetic nervous system** and the **parasympathetic nervous system**.

Sympathetic nervous system	Parasympathetic nervous system
• raises blood pressure	• reduces blood pressure
• raises the output of blood from the heart	• reduces the output of blood from the heart
• raises the rate of ventilation	• reduces the rate of ventilation
• dilates the pupil of the eye	• narrows the pupil of the eye

Comparing some effects of the sympathetic nervous system and the parasympathetic nervous system

Notice that the effects of the sympathetic nervous system oppose the effects of the parasympathetic nervous system. We say that the effects are **antagonistic**. The balance of antagonistic effects regulates the involuntary responses of muscles and glands. Control of heart rate is an example.

Fact file

Not all effects of the sympathetic and parasympathetic nervous system are antagonistic. For example, the sympathetic system stimulates sweating. However, the parasympathetic system has no comparable effect.

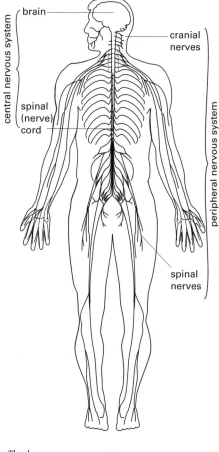

The human nervous system

The brain

Functions of the brain

The **brain** is the primary **integration** and **control organ** of the body

- receives information (**sensory input**) from sense organs in the head and from other body organs via ascending fibres of the spinal cord;
- communicates via **cranial nerves**, the **descending fibres** of the spinal cord, and through regulation of endocrine secretions via the **pituitary gland**;
- utilises many neurotransmitters, including dopamine, serotonin, histamine, GABA and endorphins;
- is able to accept sensory information from a number of sources, compare it with previous experience (learning) and make sure that the appropriate actions are initiated.

Structure of the brain

Meninges: membranes which line the skull and cover the brain. They help to protect and nourish the brain tissues, but may be infected either by a virus or a bacterium to cause the potentially fatal condition **meningitis**.

Cerebrum (cerebral cortex): has motor areas to control voluntary movement, sensory areas which interpret sensations and association areas to link the activity of motor and sensory regions. The centre of intelligence, memory, language and consciousness.

Forebrain: here many emotions are localised, and damage to this area may cause aggression, apathy, extreme sexual behaviour and other emotional disturbances.

Skull: the cranium is a bony 'box' which encloses and protects the brain.

Hypothalamus: contains centres which control thirst, hunger and thermoregulation.

Visual centre: this area of the cerebral cortex
- interprets impulses along the optic nerve i.e. responsible for vision;
- has the connector neurones for both accommodation and the pupil reflex.

Pituitary gland: is a link between the central nervous system and the endocrine system. Secretes a number of hormones, including follicle stimulating hormone which regulates development of female gametes, and anti-diuretic hormone which controls water retention by the kidney.

Cerebellum: co-ordinates movement using sensory information from position receptors in various parts of the body. Helps to maintain posture using sensory information from the inner ear. Can control learned sequences of activity involved in dancing, athletic pursuits and in the playing of musical instruments.

Medulla: the link between the spinal cord and the brain, and relays information between these two structures. Has a number of reflex centres which control:
- vital reflexes which regulate heartbeat, breathing and blood vessel diameter;
- non-vital reflexes which co-ordinate swallowing, salivation, coughing and sneezing.

Spine (vertebral column): composed of 33 separate vertebra which surround and protect the spinal cord; between each pair of vertebrae two **spinal nerves** carry sensory information into the spinal cord and motor information out of it. Dislocation of the vertebrae may compress the spinal nerves, causing great pain, or even crush the spinal cord, leading to paralysis.

Movement is brought about by skeletal muscles which pull on tendons. The tendons in turn pull on bones and cause them to move at a joint. At the elbow, the two antagonistic muscles the **biceps** and the **triceps** work together to bend and extend the elbow.

The contraction of all skeletal muscles is coordinated by the cerebrum of the brain. This part of the brain coordinates voluntary movement – it receives sensory information which is interpreted by association areas. Motor areas then send nerve impulses to bring about contraction of the desired muscle.

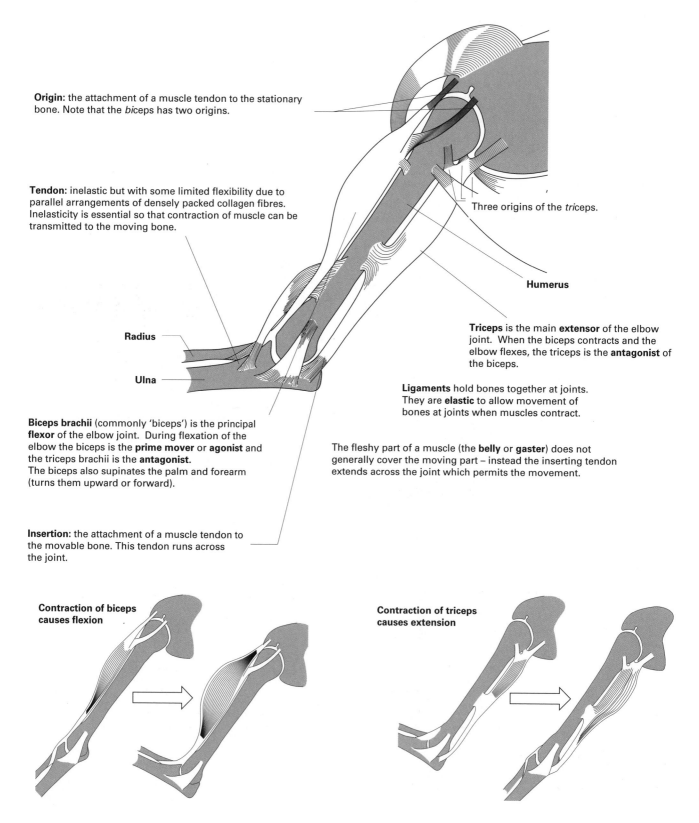

Origin: the attachment of a muscle tendon to the stationary bone. Note that the *bi*ceps has two origins.

Tendon: inelastic but with some limited flexibility due to parallel arrangements of densely packed collagen fibres. Inelasticity is essential so that contraction of muscle can be transmitted to the moving bone.

Three origins of the *tri*ceps.

Humerus

Radius

Ulna

Triceps is the main **extensor** of the elbow joint. When the biceps contracts and the elbow flexes, the triceps is the **antagonist** of the biceps.

Ligaments hold bones together at joints. They are **elastic** to allow movement of bones at joints when muscles contract.

Biceps brachii (commonly 'biceps') is the principal **flexor** of the elbow joint. During flexation of the elbow the biceps is the **prime mover** or **agonist** and the triceps brachii is the **antagonist**.
The biceps also supinates the palm and forearm (turns them upward or forward).

The fleshy part of a muscle (the **belly** or **gaster**) does not generally cover the moving part – instead the inserting tendon extends across the joint which permits the movement.

Insertion: the attachment of a muscle tendon to the movable bone. This tendon runs across the joint.

Contraction of biceps causes flexion

Contraction of triceps causes extension

When the muscles attached to the bones of the skeleton (skeletal muscle) contract (shorten) and relax (lengthen), the bones move. Looking at skeletal muscle tissue under the optical microscope shows a pattern of light and dark **striations** (bands) running from one side to another across fibre-like cells. Muscle tissue which shows a pattern of light and dark bands under the optical microscope is called **striated muscle**. Slender threads running the length of each muscle fibre are also visible. The threads are called **myofibrils**.

The electron microscope shows that

- each myofibril is made up of alternating light and dark bands
- the bands of myofibrils lying parallel next to one another line up – light to light, dark to dark – accounting for the striations running across each muscle fibre
- each myofibril consists of a number of longitudinal filaments – some thick, others thin

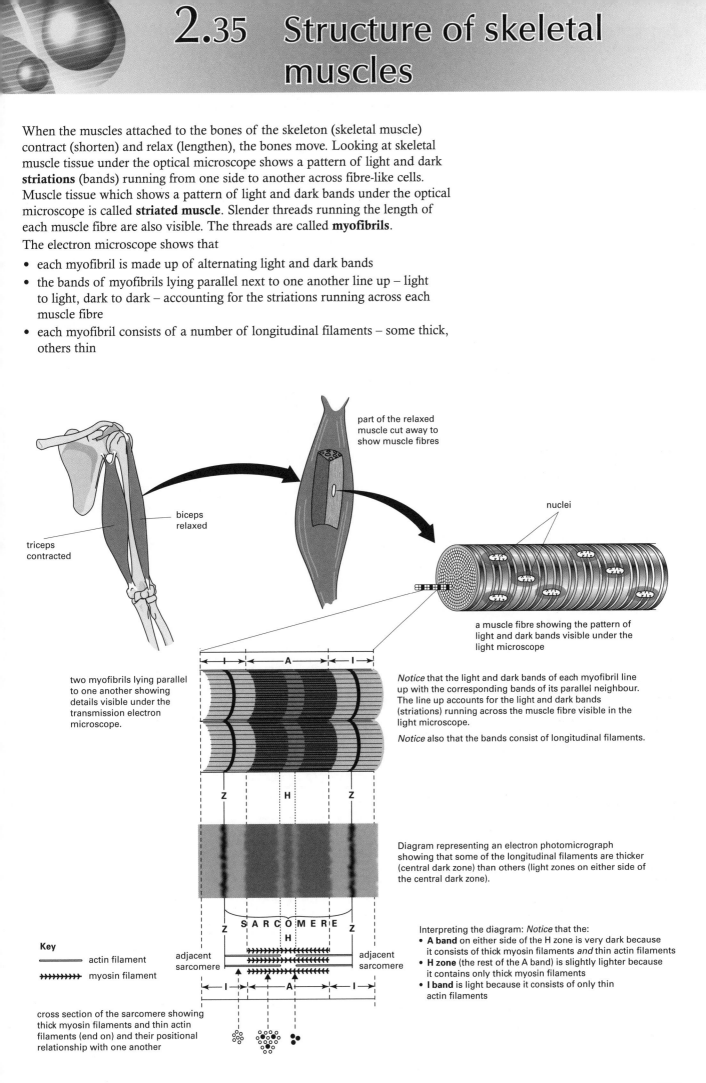

part of the relaxed muscle cut away to show muscle fibres

nuclei

biceps relaxed

triceps contracted

a muscle fibre showing the pattern of light and dark bands visible under the light microscope

two myofibrils lying parallel to one another showing details visible under the transmission electron microscope.

Notice that the light and dark bands of each myofibril line up with the corresponding bands of its parallel neighbour. The line up accounts for the light and dark bands (striations) running across the muscle fibre visible in the light microscope.

Notice also that the bands consist of longitudinal filaments.

Diagram representing an electron photomicrograph showing that some of the longitudinal filaments are thicker (central dark zone) than others (light zones on either side of the central dark zone).

Interpreting the diagram: *Notice* that the:
- **A band** on either side of the H zone is very dark because it consists of thick myosin filaments *and* thin actin filaments
- **H zone** (the rest of the A band) is slightly lighter because it contains only thick myosin filaments
- **I band** is light because it consists of only thin actin filaments

Key

—— actin filament

✦✦✦✦✦✦✦ myosin filament

adjacent sarcomere

adjacent sarcomere

cross section of the sarcomere showing thick myosin filaments and thin actin filaments (end on) and their positional relationship with one another

119

Remember that

- thick filaments consist of the protein **myosin**
- thin filaments consist of the protein **actin**
- two other proteins called **tropomyosin** and **troponin** are bound to actin
- the enzyme **ATP-ase** is bound to myosin
- the arrangement of the filaments corresponds to the pattern of the light and dark bands of muscle tissue seen under the optical microscope

Notice that the thin actin filaments are anchored to a dark band called the **Z line** running through the middle of each light I band. The thick myosin filaments are anchored to the **M line** (not shown) running through the middle of the H zone. The region between one Z line and the next is called the **sarcomere**. It is the functional unit of muscle tissue.

Muscle fibres and the T-system

Remember that skeletal muscle tissue is made up of many fibre-like cells. Each fibre is:

- filled with cytoplasm (called **sarcoplasm**) containing many nuclei
- surrounded by the **sarcolemma** which is similar to the plasma membrane surrounding cells

… *except* that the sarcolemma folds inwards forming a series of transverse tubules which run through the sarcoplasm. The tubules are the components of the **T-system**.

The endoplasmic reticulum of a muscle fibre is called the **sarcoplasmic reticulum**. It forms swollen vesicles at the Z lines of the sarcomeres. Here the vesicles are in contact with the tubules of the T-system. The vesicles contain a high concentration of calcium ions (Ca^{2+}). The T-system and sarcoplasmic reticulum surround the myofibrils of the muscle fibre.

The diagram is a simplified representation of the arrangement.

Questions

1 What are myofibrils?

2 Draw a labelled diagram that identifies the pattern of light and dark bands of myofibrils.

3 What is a sarcomere?

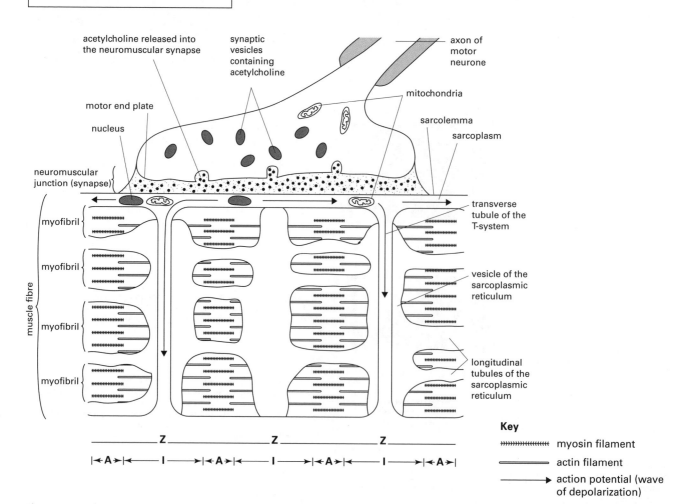

The structure of part of a muscle fibre and the neuromuscular junction

The neuromuscular junction

Nerves communicate with each other at synapses. To bring about muscle contraction, nerves communicate with muscles at **neuromuscular junctions**.

Synapse	Neuromuscular junction
Gap (synaptic cleft) between two or more neurones	Gap (synaptic cleft) between motor neurone and the **end plate** of muscle fibre – a region where the muscle fibre membrane is highly folded
Synapse contains many mitochondria and synaptic vesicles.	Neuromuscular junction contains many mitochondria and synaptic vesicles.
Nerve impulse is transmitted by neurotransmitter such as acetylcholine released from synaptic vesicles across the synaptic cleft. This release is stimulated by calcium ions.	Nerve impulse is transmitted by acetylcholine released from synaptic vesicles across the synaptic cleft. This release is stimulated by calcium ions.
End result is **action potential** propagated along postsynaptic fibre.	End result is an **end plate potential**. End plate potentials add together to stimulate an action potential which spreads trough **T tubules** and leads to muscle contraction.
Neurotransmitter broken down by enzymes once it has had effect.	Neurotransmitter broken down by enzymes once it has had effect.

The synapse and neuromuscular junction compared

Other types of muscle

The muscle fibres that bring about movement of bones at joints under control of the cerebrum are **skeletal muscle**, also called **striated muscle** or **voluntary muscle**. 'Striated' refers to their striped appearance under the microscope, which results from the arrangement of actin and myosin fibres in the sarcomeres.

Smooth muscle or **involuntary muscle** occurs in the walls of blood vessels and in the gut. This muscle lacks the striped appearance of voluntary muscle. It is under the control of the autonomic nervous system.

Cardiac muscle is found in the heart. It is striated, but its structure differs from that of voluntary muscle in that the fibres are cross-connected. It is **myogenic** – it contracts spontaneously and is not subject to fatigue as is voluntary muscle.

Nervous and hormonal control

An individual's coordinated response to changes in the environment (stimuli) is the result of coordinated activities of the nervous system and endocrine system.

The responses to the hormone adrenaline are fast and short lived – they mimic the action of the sympathetic nervous system. They result in an increased supply of oxygenated blood containing glucose to the voluntary muscles, ready for fight or flight. They also increase the level of mental awareness so that decisions can be taken about the most appropriate response to the danger.

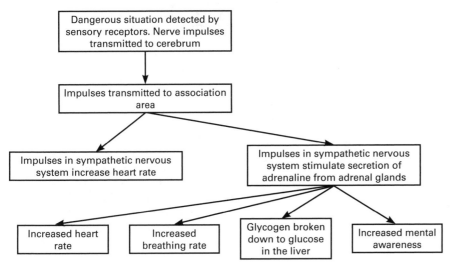

The fight or flight response is both nervous and hormonal

A myofibril consists of a chain of sarcomeres. Understanding how a sarcomere contracts is the key to understanding how a myofibril contracts and therefore the contraction of muscle tissue as a whole.

How does muscle contract?

In the early 1950s, research groups at Cambridge University and University College London tackled the problem. Different types of microscopy provided the evidence.

- Electron microscopy showed that actin and myosin filaments remained the same length whether a sarcomere (and therefore a myofibril) was contracted or relaxed.

- Phase contrast microscopy showed that the pattern of light and dark bands changed during the contraction and relaxation of a muscle fibre.

Combined, the evidence strongly suggested that the filaments of actin and myosin slide past one another as the length of the sarcomere changes – long when relaxed; short when contracted. In 1954 both research groups proposed the sliding filament theory of muscle contraction to explain their observations.

The **phase contrast microscope** alters the properties of light passing through it. This causes a difference in brightness in the cell's structures, improving contrast. The improvement makes it possible to see the structures one from another.

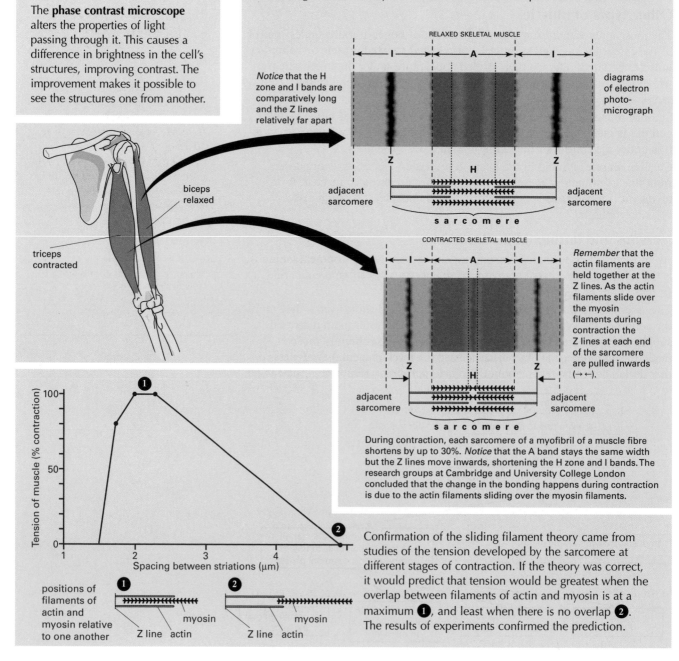

Notice that the H zone and I bands are comparatively long and the Z lines relatively far apart

diagrams of electron photo-micrograph

Remember that the actin filaments are held together at the Z lines. As the actin filaments slide over the myosin filaments during contraction the Z lines at each end of the sarcomere are pulled inwards (→ ←).

During contraction, each sarcomere of a myofibril of a muscle fibre shortens by up to 30%. *Notice* that the A band stays the same width but the Z lines move inwards, shortening the H zone and I bands. The research groups at Cambridge and University College London concluded that the change in the bonding happens during contraction is due to the actin filaments sliding over the myosin filaments.

Confirmation of the sliding filament theory came from studies of the tension developed by the sarcomere at different stages of contraction. If the theory was correct, it would predict that tension would be greatest when the overlap between filaments of actin and myosin is at a maximum ❶, and least when there is no overlap ❷. The results of experiments confirmed the prediction.

positions of filaments of actin and myosin relative to one another

Contraction

How actin and myosin filaments slide past one another

Electron microscopy reveals cross bridges between filaments of actin and myosin. X-ray crystallography shows that the cross bridges move during contraction.

When muscle is relaxed, the myosin heads are held away from the myosin binding sites on the actin filaments. When muscle is contracted, the myosin heads move out from their resting position and link to the myosin binding sites on the actin filaments, forming myosin/actin cross-bridges. The bridges stand at 45° from the myosin filaments when contraction begins.

Remember:

- Tropomyosin and troponin are proteins bound to actin. The diagram on the next page shows the arrangement.
- Muscle tissue shows ATP-ase activity. The enzyme is located in each myosin head to which ADP and P_i are also bound when the muscle fibre is relaxed.
- Nerve impulses cause muscles to contract.

Contraction of muscle depends on the interactions between the T-system, sarcoplasmic reticulum, and actin, myosin and the proteins bound to them.

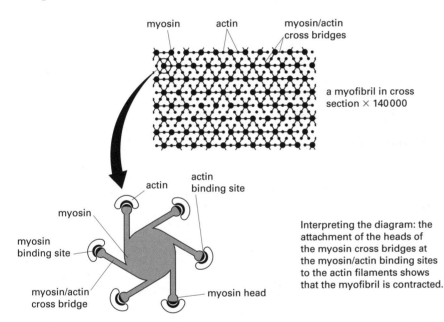

myosin   actin   myosin/actin cross bridges

a myofibril in cross section × 140000

actin

actin binding site

myosin

myosin binding site

myosin/actin cross bridge

myosin head

Interpreting the diagram: the attachment of the heads of the myosin cross bridges at the myosin/actin binding sites to the actin filaments shows that the myofibril is contracted.

Myosin/actin cross-bridges project outwards from the myosin at regular intervals of 6–7 nm. Each bridge is separated by 60° from its neighbour and stands at 45° from the myosin when contraction begins.

When a nerve impulse travelling along a motor neurone arrives at the **neuromuscular junction** (the synapse between the motor neurone and the sarcolemma of the muscle fibre) the **motor end plate** (a specialized part of sarcolemma) is depolarized.

- As a result a wave of depolarization travels along the sarcolemma and passes through the T-system into the muscle fibre.
- As a result the membranes of the vesicles of the sarcoplasmic reticulum become permeable to the calcium ions (Ca^{2+}) stored within them.
- As a result the calcium ions diffuse down their concentration gradient out of the vesicles and into the sarcoplasm surrounding the myofibrils.
- As a result the sites where myosin attaches to actin are uncovered and myosin/actin cross-bridges form, linking the filaments of actin and myosin.
- As a result the actin filaments are pulled over by the myosin towards the centre of the sarcomere.

With all the myofibrils of a sarcomere acting in this way, a force is generated which leads to a contraction (shortening) of the sarcomere (and therefore the myofibril, muscle fibre, and muscle tissue as a whole).

The diagram shows you the sequence of events.

Myosin moves actin during the contraction of muscle. The dashed line is a marker showing that contraction of a myofibril occurs as a result of the movement of actin filaments sliding over filaments of myosin. The overlap of actin and myosin filaments accounts for the changes in the pattern of light and dark bands of a sarcomere (and therefore a myofibril and a muscle fibre) during contraction of muscle from its relaxed state.

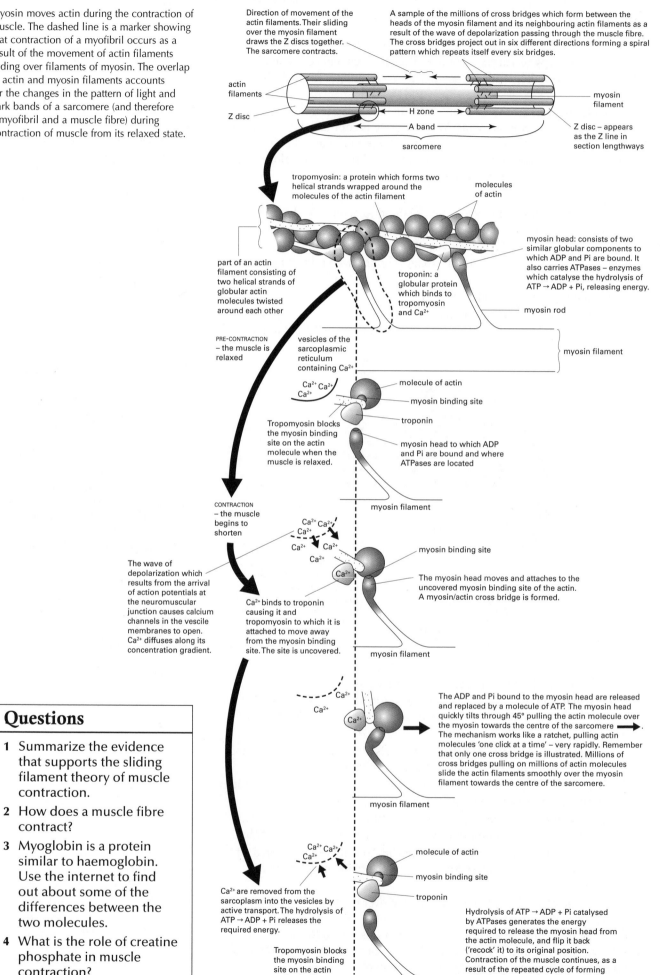

Direction of movement of the actin filaments. Their sliding over the myosin filament draws the Z discs together. The sarcomere contracts.

A sample of the millions of cross bridges which form between the heads of the myosin filament and its neighbouring actin filaments as a result of the wave of depolarization passing through the muscle fibre. The cross bridges project out in six different directions forming a spiral pattern which repeats itself every six bridges.

actin filaments

Z disc

myosin filament

Z disc – appears as the Z line in section lengthways

H zone

A band

sarcomere

tropomyosin: a protein which forms two helical strands wrapped around the molecules of the actin filament

molecules of actin

myosin head: consists of two similar globular components to which ADP and Pi are bound. It also carries ATPases – enzymes which catalyse the hydrolysis of ATP → ADP + Pi, releasing energy.

part of an actin filament consisting of two helical strands of globular actin molecules twisted around each other

troponin: a globular protein which binds to tropomyosin and Ca^{2+}

myosin rod

myosin filament

PRE-CONTRACTION – the muscle is relaxed

vesicles of the sarcoplasmic reticulum containing Ca^{2+}

Ca^{2+} Ca^{2+} Ca^{2+}

molecule of actin

myosin binding site

troponin

Tropomyosin blocks the myosin binding site on the actin molecule when the muscle is relaxed.

myosin head to which ADP and Pi are bound and where ATPases are located

myosin filament

CONTRACTION – the muscle begins to shorten

Ca^{2+} Ca^{2+} Ca^{2+} Ca^{2+} Ca^{2+} Ca^{2+}

myosin binding site

The myosin head moves and attaches to the uncovered myosin binding site of the actin. A myosin/actin cross bridge is formed.

The wave of depolarization which results from the arrival of action potentials at the neuromuscular junction causes calcium channels in the vescile membranes to open. Ca^{2+} diffuses along its concentration gradient.

Ca^{2+} binds to troponin causing it and tropomyosin to which it is attached to move away from the myosin binding site. The site is uncovered.

myosin filament

Ca^{2+}

Ca^{2+}

The ADP and Pi bound to the myosin head are released and replaced by a molecule of ATP. The myosin head quickly tilts through 45° pulling the actin molecule over the myosin towards the centre of the sarcomere ⟶. The mechanism works like a ratchet, pulling actin molecules 'one click at a time' – very rapidly. Remember that only one cross bridge is illustrated. Millions of cross bridges pulling on millions of actin molecules slide the actin filaments smoothly over the myosin filament towards the centre of the sarcomere.

myosin filament

Questions

1 Summarize the evidence that supports the sliding filament theory of muscle contraction.

2 How does a muscle fibre contract?

3 Myoglobin is a protein similar to haemoglobin. Use the internet to find out about some of the differences between the two molecules.

4 What is the role of creatine phosphate in muscle contraction?

Ca^{2+} Ca^{2+} Ca^{2+}

molecule of actin

myosin binding site

troponin

Ca^{2+} are removed from the sarcoplasm into the vesicles by active transport. The hydrolysis of ATP → ADP + Pi releases the required energy.

Tropomyosin blocks the myosin binding site on the actin molecule. The muscle is relaxed.

Hydrolysis of ATP → ADP + Pi catalysed by ATPases generates the energy required to release the myosin head from the actin molecule, and flip it back ('recock' it) to its original position. Contraction of the muscle continues, as a result of the repeated cycle of forming and releasing myosin/actin cross bridges.

myosin filament

The word **behaviour** refers to the responses of animals to stimuli.

- **Innate behaviour** is behaviour that is not learned – it is an automatic response that has been determined genetically. An example is various **escape reflexes** in response to danger seen in animals such as the giant squid.
- **Learned behaviour** is built up based on previous experience.

Innate behaviour

Taxes and **kineses** are simple forms of innate behaviour. The terms describe the movements of animals (and other mobile organisms) that help them find environments where they are most likely to survive.

- Taxes (*singular* taxis) are **directional** movements which orientate an individual with respect to the stimulus causing the response.
  - **Positive (+) taxis** – an individual moves *towards* the stimulus.
  - **Negative (–) taxis** – an individual moves *away* from the stimulus.

Tactic responses are classified according to the type of stimulus, e.g. light – phototaxis; chemicals – chemotaxis.

- Kineses (*singular* kinesis) are random **non-directional** movements: an individual does not move towards or away from a stimulus but
  - moves faster
  - changes direction more frequently

  … in response to the intensity of a stimulus which threatens its survival. The more intense the threatening stimulus, the faster are the movements and the more frequent the changes of direction.

  In this way the individual is more likely to find quickly a less threatening environment. The movements then slow and may stop altogether, and so the individual spends more time in an environment where it is more likely to survive. Like taxes, kinetic responses are classified according to the type of stimulus, e.g. photokinesis.

Examples

- Maggots (fly larvae) and woodlice quickly lose water from the body in dry air, threatening their survival. The air of open, brightly lit habitats is likely to be drier than the air of shaded, dimly lit habitats.
  - Ⓡ As a result maggots and woodlice are more likely to be found in dimly lit habitats where air is **humid** (saturated with water vapour).
  - Ⓡ As a result, loss of water from the body is reduced.
  - Ⓡ As a result maggots and woodlice are more likely to survive.

How do maggots and woodlice find dimly lit habitats?

- Maggots are negatively phototactic: they move away from bright light.
  - Ⓡ As a result they move towards dim light.
  - Ⓡ As a result they are more likely to survive in the humid air associated with dimly lit habitats.

- Woodlice are photokinetic: they move faster and change direction more frequently the brighter (more intense) the light (and therefore the drier the air).
  - Ⓡ As a result, the more active woodlice are more likely to find a dimly lit habitat where the air is more likely to be humid.
- Woodlice are much less active in dimly lit habitats.
  - Ⓡ As a result they are more likely to remain in dimly lit habitats and survive.

Investigating photokinesis

You can use a **choice chamber** to investigate the responses of woodlice to dry/humid and light/dark environments. The diagram gives you the idea. Remember that the responses of woodlice are an example of photokinesis.

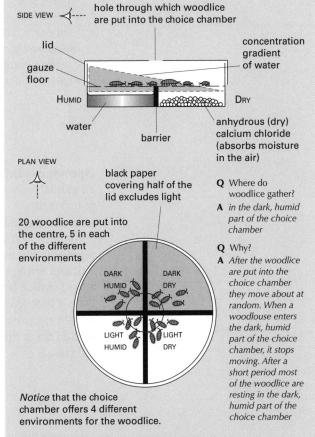

Q Where do woodlice gather?

A *in the dark, humid part of the choice chamber*

Q Why?

A *After the woodlice are put into the choice chamber they move about at random. When a woodlouse enters the dark, humid part of the choice chamber, it stops moving. After a short period most of the woodlice are resting in the dark, humid part of the choice chamber*

Notice that the choice chamber offers 4 different environments for the woodlice.

The choice chamber allows woodlice to find and remain in an environment which favours their survival.

Questions

1. Explain the difference between taxes and kineses.
2. Briefly summarize how maggots and woodlice find habitats where they are more likely to survive.
3. Describe how you set up an investigation into the responses of woodlice to dry/humid and light/dark environments.

Learned behaviour

Learned behaviour involves an adaptive change in response based on previous experience.

Habituation occurs when an **insignificant stimulus** (i.e. one that is neither **beneficial** nor **harmful**) is repeated, and an animal learns **not** to respond to it,

e.g. not responding to background traffic noise, but still waking up when an alarm clock rings.

The advantage is that, by ignoring these stimuli, animals spend both time and energy more efficiently.

Classical conditioning occurs when an animal learns to associate a **neutral** stimulus with an important one: the response to the important stimulus is now a **conditioned response**, and is

- shown in the presence of the neutral stimulus (called the **conditioned** stimulus)
- involuntary, and reinforced by repetition
- temporary, and will be **unlearned** if the neutral stimulus completely replaces the important one.

Pavlov: showed dogs would salivate on hearing a bell normally rung when receiving food.

Operant conditioning ('trial and error' learning) occurs when an animal learns to associate a response with a reward or a punishment.

Skinner: used a specially-devised box (a Skinner box) which allowed rats or pigeons to choose an action – through trial and error they learned which action (which button to press) gave them a reward of food.

A mild electric shock could also be used to teach them which button not to press.

mmm...
... which one?

RATTY GUMS

Correct behaviour reward

Incorrect behaviour: punishment

Humans use 'trial and error' learning: a golfer might try a different grip on the club when hitting a golf shot.

Learned behaviour modified by experience

Latent learning occurs when an animal acquires knowledge at a certain time, without any positive reinforcement, but does not use it until a later time when the knowledge is needed.

Example: learning to use a screwdriver by watching a parent, without any obvious reward at the time

Example: learning a route when being driven to school.

Insight behaviour (reasoning) is the most complex form of learning:

• is the ability to solve a problem without trial and error – may involve recall of past experiences.

Köhler: showed that chimpanzees could 'work out' (reason) how to reach fruit. Only humans and other primates can do this.

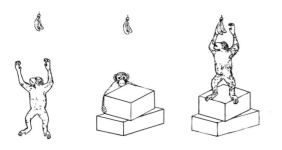

Imprinting involves:

• a combination of innate **and** learned behaviour
• occurs within a limited **critical period** of birth

Lorenz: showed that ducklings would follow a human if the human was present as the duckling hatched. There is **innate** behaviour to follow the mother, and **learned** behaviour about who the mother is.

Behaviour in primates

What about human social behaviour?
- form social groups which may extend beyond survival e.g. based on income, nationality or cultural interests
- have a highly-developed power of speech, and can use written and drawn symbols for communication
- have an extended range due to their ability to modify their environment – make homes in fixed position
- make complex tools, and have learned to control fire for cooking and heating.

Mothers and infants: form the basic social group for many primates
- mother-infant bonding needed for young to develop into socially-capable adults
- may continue after infancy, with females remaining together and males dispersing to other groups
- combined power of such a female group may mean they are dominant to the alpha-male.

Social behaviour

Dominance: primates are mainly group-living animals and form **dominance hierarchies.**

| higher-ranked animals displace lower-ranked animals from food, mates and 'space'. They usually have more reproductive success (often through mating more frequently) | are **not** fixed, and depend on age, sex, aggression and intelligence

rank is learned through play, fighting and peer support behaviour. |

Grooming: grooming of others (allogrooming) is an important support mechanism:
- subordinates groom dominant animals
- males groom females for sexual access
- mothers groom infants to clean their fur.

Communication
- includes scents, vocalisation, postures and gestures
- may be reflexes indicating emotional state e.g. teeth baring/eyelid flashing or may have a more specific purpose e.g. loud territorial calls in howler monkeys
- can be used to reduce aggression e.g. presentation of rear, and offers of copulation in baboons.

There has been some success in teaching sign language to chimpanzees!

Aggression
- may occur **within** the group to keep dominance hierarchies
- may involve communal activity to **protect** the group e.g. baboons cooperate to repel leopards
- some scientists suggest that bipedalism evolved so males could look taller and could beat their chests during displays of aggression.

Reproduction: all primate females (except for Humans) show a seasonal or cyclical willingness to mate.
- Female usually shows visual changes (e.g. swellings around the anus and/or vulva) so it is clear when she is ready to mate.
- Bonobos (pygmy chimps) may have periods of frequent copulation which may be important in pair bonding.

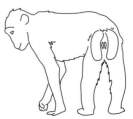

Cultural behaviour: this is learned behaviour that is passed from generation to generation.
e.g.
- Japanese Macaques: learned that sweet potatoes taste better if sand is washed off.
- Chimpanzees: 'termite-fishing' using pointed sticks seems to be learned by all members of a group, and passed on to infants.
- Many examples in Humans, including use of writing and drawing instruments.

The DRD4 dopamine receptor and human behaviour

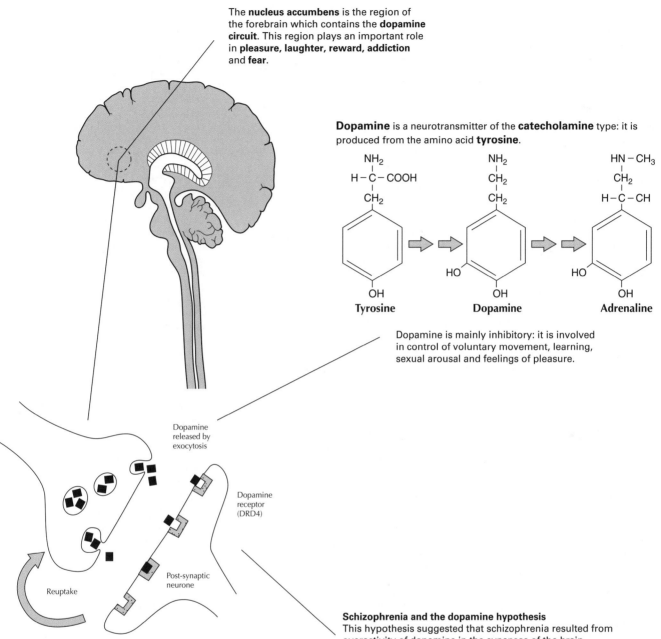

The **nucleus accumbens** is the region of the forebrain which contains the **dopamine circuit**. This region plays an important role in **pleasure, laughter, reward, addiction** and **fear**.

Dopamine is a neurotransmitter of the **catecholamine** type: it is produced from the amino acid **tyrosine**.

Tyrosine   Dopamine   Adrenaline

Dopamine is mainly inhibitory: it is involved in control of voluntary movement, learning, sexual arousal and feelings of pleasure.

Dopamine released by exocytosis

Dopamine receptor (DRD4)

Post-synaptic neurone

Reuptake

Parkinson's Disease results from a shortage of dopamine in the **basal ganglia** of the brain.
- Symptoms include rigidity of muscles, tremor (shaking) when resting, slowness of movement and a difficulty in starting voluntary activity.
- It can be treated with L-dopa, a compound which can be converted to dopamine. This is a test for the disease – symptoms are less severe when the drug is used.

Stem cells may help!
Stem cells injected into the basal ganglia may differentiate and produce new dopamine-producing cells. This may overcome symptoms of Parkinson's Disease.

Cocaine and DRD4
- Dopamine stimulates 'pleasure centres'
- Cocaine occupies the transporter molecules which remove dopamine from the synapse and return it to the pre-synaptic neurone.
- Cocaine therefore amplifies the dopamine response – pleasure and arousal.
- Addiction develops as an attempt to avoid dopamine depletion.

Schizophrenia and the dopamine hypothesis
This hypothesis suggested that schizophrenia resulted from overactivity of dopamine in the synapses of the brain. Evidence included:
- anti-schizophrenic drugs work by blocking post-synaptic dopamine receptors. Overdosing with these drugs causes symptoms similar to Parkinson's Disease;
- post-mortem examination of the brains of schizophrenics show increased amounts of dopamine and more DRD4 sites;
- large doses of amphetamines (which increase dopamine activity) create symptoms similar to schizophrenia.

How do we know?
Mice can be genetically-engineered to mutate the DRD4 gene.
- Some have learning disorders (similar to ADHD)
- Many lack ability to become aroused.
- Some develop hypertension (dopamine may help to control salt release in the urine).

Index

ribulose bisphosphate (RuBP) 43, 45
ribulose bisphosphate carboxylase (rubisco) 43
rice 92
RNA 57–9
  interference 59
  polymerase 58, 62–3
  see also mRNA, rRNA, small interfering RNA, tRNA
RNA induced silencing complex (RISC) 59
rods 13
roots 12
rRNA 57

saltatory conduction 17–18
sampling 104–5
sarcolemma 120, 123–4
sarcomere 119–20, 122–4
sarcoplasm 120
sarcoplasmic reticulum 120, 123–4
saturation 44
savanna 109
schizophrenia 129
Schwann cell 14
second messenger 6, 23, 25
secondary consumers 96
secondary metabolites 84
secondary sexual characteristics 22–3
secondary succession 103
secretion 22, 26, 31, 58–9
selection pressure 76
selective reabsorption 34
semi-desert 109
sensitivity 116
sensory input 117
sensory neurone 12–13, 28–9
sensory receptors 12–13
sere 102
serial dilution 85
serotonin 117
sessile species 105
set point 9
sex chromosomes 68
sex-inked inheritance 68
shade plants 45
shoots 12
sickle cell anaemia 75
sigmoid growth curve 107
signal molecule specificity 6
silencing genes 59
silent mutation 60
sino-atrial node (SAN) 28–9
sinusoid 31–2
skeletal muscle 11–12, 118–24
skin 11
skull 117
sliding filament theory 122–4
small interfering RNA (siRNA) 59
smooth muscle 121
snails 105
snapdragons 67–8
social behaviour 128
sodium ions Na$^+$ 13–20, 23, 34, 47
sodium/potassium pump 15, 16–18
soil 101
somatic line therapy 93
somatic nuclear transfer 80
speciation 73, 76–7
species 73, 74, 76–7, 78, 86, 95
sperm mother cell 66
spinal column/cord 116–17
spinal nerves 116–17
spliceosome 59

sports drinks 35
squirrels 110
stabilizing selection 76–7
stalked particles 53
standing crop 97
starch grains 38
stationary phase 81
stem cell therapy 27
stem cells 27, 79–80, 129
steroid abuse, testing 30
sticky ends 87, 88, 91
stimuli 8–9, 11, 12–13, 16, 26
  directional 12
stop codon 59
streak pattern 85
striated muscle 118–24
stroma 38–9
structural gene 62–3
sub-species 77
substrate 48
substrate–level phosphorylation 48, 50, 53
subtilisin 82
succession 102–3
summation 17
sun plants 45
surrogate mothers 78, 80
survival 12
sweat 35
sweat glands 11
sweep net 104–5
sympathetic nervous system 28–9, 31, 116, 121
synapse 19–21, 120–1
synaptic cleft 19–21
synaptic knob 19–23
synaptic transmission 19–21
synaptic vesicles 19–21, 120

target cell 6–7, 23
target tissue 6–7, 22, 25
taxes 125
telophase 65
temperature
  and biotechnology 81, 84
  control 8–9, 10–11
  and enzyme–controlled reactions 61, 71
  optimum 8
  and impulse conduction 18
  and populations 108–9, 112
  and rate of photosynthesis 44–6
  and variation 71
  see also body temperature
temporal summation 20
tendon 118
terminal transferase 88
termination 58–9
tertiary consumers 96
testes 22, 65
testosterone 22, 30
tetanus 21
thermoreceptor 11, 13
thermoregulatory centre 11
threshold value 17–18
thylakoid membranes 38–9, 41–2
thylakoid space 41
thymine (T) 56
thyroid gland 22
thyroxine 11, 22
Ti plasmid 88–9
tissue culture 79
tissue development 64
tissue fluid 8, 35
tortoises 112

total count 85
totipotent cells 27
touch 13
toxins 30, 36, 81
transcription 58–9, 62–3
transcription factors 62–3, 64
transducer 13
transformed organisms 88–9
translation 58–9, 62–3
transplantation 27, 92
transporter proteins 31
tree beating 104
trial and error learning 126
tricarboxylic acid cycle 53
triceps 118–9, 122
triose phosphate (TP) 43, 45
Triticum aestivum 78
tRNA 57–9
trophic levels 96–7
tropical forest 109
tropisms 12, 113
tropomyosin 120, 122–4
troponin 120, 122–4
T-system 120–1, 123
tubular secretion 34
tumour 61
tumour suppressor genes 61
type 1 diabetes 25, 27, 91
type 2 diabetes 25
tyrosine 129

ulna 118
ultrafiltration 34
uncouplers 55
urea 30, 32, 36–7
ureter 33–4, 37
uric acid 30
urinary system 33–7
urine 10, 33–7
  testing 30, 82

vagus nerve 28–9
variation 70–7
  environmental 70–1
  genetic 70–1, 72–3
vasoconstriction 11
vasodilation 11
vector 88–93
vegetative propagation 79
ventral surface 64
viable count 85
visual centre 117
vitamin A 32, 40, 92
vitamins 32
voltage gated channels 16–18, 19
voluntary muscle 118–4
voluntary response 116

water
  balance 35
  and photosynthesis 39, 42, 44
  potential 8, 35
Weinberg, W 74
woodland 96–7, 104
woodlice 109, 125

xenotransplantation 92

yeast 49–51

Z line 120, 122–4
zygote 64, 66, 71